AP STATISTICS
ALL ACCESS

Robin Levine-Wissing
Instructional Supervisor for Mathematics
and AP Statistics Instructor
Glenbrook North High School
Northbrook, Illinois

David W. Thiel
Mathematics Specialist
Southern Nevada Regional Professional
Development Program
Adjunct Professor, Department of Economics
University of Nevada–Las Vegas
Las Vegas, Nevada

Research & Education Association
Visit our website: www.rea.com/studycenter

Research & Education Association

61 Ethel Road West
Piscataway, New Jersey 08854
E-mail: info@rea.com

AP STATISTICS ALL ACCESS

Library of Congress Control Number 2011943735

ISBN-13: 978-0-7386-1058-0
ISBN-10: 0-7386-1058-5

Contents

About Our Authors vi

About Research & Education Association vii

Acknowledgments vii

Chapter 1: Welcome to REA's All Access for AP Statistics   1

Chapter 2: Strategies for the Exam   7

What Will I See on the AP Statistics Exam? 7

Section I: Strategies for the Multiple-Choice Section of the Exam 9

Section II: Strategies for the Free-Response Section of the Exam 16

Chapter 3: Exploring Data   25

Constructing and Interpreting Graphical Displays of
Distributions of Univariate Data 25

Summarizing Distributions of Univariate Data 37

Comparing Distributions of Univariate Data (Dotplots, Back-to-Back
Stemplots, Parallel Boxplots) 49

Exploring Bivariate Data 52

Exploring Categorical Data 70

Quizzes 1 and 2 .... *www.rea.com/studycenter*

Chapter 4: Sampling and Experimentation 83

Overview of Methods of Data Collection 83

Planning and Conducting Surveys 85

Planning and Conducting Experiments 94

Generalizability of Results and Types of Conclusions that can be
Drawn from Observational Studies, Experiments, and Surveys 100

Quizzes 3 and 4 .... *www.rea.com/studycenter*

Mini-Test 1 .... *www.rea.com/studycenter*

Chapter 5: Anticipating Patterns 103

Probability 103

Combining Independent Random Variables 129

The Normal Distribution 132

Sampling Distributions 145

Quizzes 5 and 6 .... *www.rea.com/studycenter*

Chapter 6: Statistical Inference 173

Estimation (Point Estimators and Confidence Intervals) 173

Tests of Significance 197

Quizzes 7 and 8 .... *www.rea.com/studycenter*

Mini-Test 2 .... *www.rea.com/studycenter*

Practice Exam (also available online at *www.rea.com/studycenter*) 247

Answer Key 275

Detailed Explanations of Answers 276

Answer Sheet 288

Appendix A 291

Glossary 301

Index 309

About Our Authors

Robin Levine-Wissing is currently the Instructional Supervisor for Mathematics and an AP Statistics Instructor at Glenbrook North High School in Northbrook, Illinois. She holds a bachelor's degree in Mathematics and Education from American University, in Washington, D.C., a master's degree in Communication Sciences from Kean University of New Jersey, and an Educational Leadership certificate from North Central College in Naperville, Illinois.

She has been teaching mathematics for 28 years, AP Statistics for 10 years, and has been a reader of the AP Statistics exams for 6 years. In June 2006, Robin became a table leader for the College Board. She has been an AP-College Board Faculty Consultant and workshop presenter since 2000. Robin is also currently Teachers Teaching with Technology (T^3) National Instructor, training teachers on handheld and computer technology for mathematics classes. She has presented workshops across the continental United States, Hawaii, and Canada for teachers of mathematics and statistics.

Robin was the 1993 recipient of the Presidential Award for Excellence in Mathematics Teaching, a 1998 Tandy Technology Scholar, and 1996 Clark County Teacher of the Year.

Robin lives with her husband, Ed, and dog, Sydney, in suburban Chicago, Illinois. They enjoy travel and foreign films.

—I wish to thank my wonderful husband, Ed, for all of his support and encouragement with this project. I could not have completed this work without you by my side. Thank you, Sydney, my dog, who often jumped up on the desk chair and sat on my lap while I was typing away, working on this book. Lastly, to Rocky and Lightning, I miss you more and more each day. I will always love you.

David Thiel is a 20-year teacher of mathematics and physics in Las Vegas, Nevada, and has a passion for probability and statistics. He has been teaching statistics for 15 years and has been involved in the Advanced Placement Statistics Program since its inception in 1996. He is a College Board consultant for AP Statistics and has been an exam reader for 6 years.

David presently spends his days as a K-12 mathematics specialist for the Southern Nevada Regional Professional Development Program, where he provides training to teachers of all levels in mathematics content and pedagogy. David is also an adjunct professor in the Department of Economics and the University of Nevada—Las Vegas, where he teaches courses in introductory statistics.

David has earned several honors in his teaching career, including the Presidential Award for Excellence in Mathematics Teaching, the Tandy Technology Scholars Award, and Outstanding Part-time Instructor at UNLV. He is a graduate of Montana State University, from which he holds both a B.S. and an M.S. in Mathematics.

David and his wife have one son and one Labrador retriever. He enjoys playing strategy games and watching baseball. He believes there is no better place in the world to study probability than Las Vegas.

—*For Kim, Alex, and Franklin. Thank you for all of your patience and support.*

—*Special thanks to Steve Goodman of Glenbrook North High School for his assistance in scrutinizing the practice exams and in creating their solutions.*

About Research & Education Association

Founded in 1959, Research & Education Association is dedicated to publishing the finest and most effective educational materials—including software, study guides, and test preps—for students in middle school, high school, college, graduate school, and beyond. Today, REA's wide-ranging catalog is a leading resource for teachers, students, and professionals.

Acknowledgments

REA would like to thank Larry B. Kling, Vice President, Editorial, for supervising development; Pam Weston, Publisher, for setting the quality standards for production integrity and managing the publication to completion; John Paul Cording, Vice President, Technology, for coordinating the design and development of the REA Study Center; Diane Goldschmidt and Michael Reynolds, Managing Editors, for coordinating development of this edition; Claudia Petrilli, Graphic Designer, for interior book design; S4Carlisle Publishing Services for typesetting; and Weymouth Design and Christine Saul for cover design.

Welcome to REA's All Access for AP Statistics

A new, more effective way to prepare for your AP exam.

There are many different ways to prepare for an AP exam. What's best for you depends on how much time you have to study and how comfortable you are with the subject matter. To score your highest, you need a system that can be customized to fit you: your schedule, your learning style, and your current level of knowledge.

This book, and the free online tools that come with it, will help you personalize your AP prep by testing your understanding, pinpointing your weaknesses, and delivering flashcard study materials unique to you.

Let's get started and see how this system works.

How to Use REA's AP All Access

The REA AP All Access system allows you to create a personalized study plan through three simple steps: targeted review of exam content, assessment of your knowledge, and focused study in the topics where you need the most help.

Here's how it works:

Review the Book	Study the topics tested on the AP exam and learn proven strategies that will help you tackle any question you may see on test day.
Test Yourself & Get Feedback	As you review the book, test yourself. Score reports from your free online tests and quizzes give you a fast way to pinpoint what you really know and what you should spend more time studying.
Improve Your Score	Armed with your score reports, you can personalize your study plan. Review the parts of the book where you are weakest, and use the REA Study Center to create your own unique e-flashcards, adding to the 100 free cards included with this book.

Finding Your Weaknesses: The REA Study Center

The best way to personalize your study plan and truly focus on your weaknesses is to get frequent feedback on what you know and what you don't. At the online REA Study Center, you can access three types of assessment: topic-level quizzes, mini-tests, and a full-length practice test. Each of these tools provides true-to-format questions and delivers a detailed score report that follows the topics set by the College Board.

Topic-Level Quizzes

Short, 15-minute online quizzes are available throughout the review and are designed to test your immediate grasp of the topics just covered.

Mini-Tests

Two online mini-tests cover what you've studied in each half of the book. These tests are like the actual AP exam, only shorter, and will help you evaluate your overall understanding of the subject.

Full-Length Practice Test

After you've finished reviewing the book, take our full-length exam to practice under test-day conditions. Available both in this book and online, this test gives you the most complete picture of your strengths and weaknesses. We strongly recommend that you take the online version of the exam for the added benefits of timed testing, automatic scoring, and a detailed score report.

Improving Your Score: e-Flashcards

Once you get your score report, you'll be able to see exactly which topics you need to review. Use this information to create your own flashcards for the areas where you are weak. And, because you will create these flashcards through the REA Study Center, you'll be able to access them from any computer or smartphone.

Not quite sure what to put on your flashcards? Start with the 100 free cards included when you buy this book.

After the Full-Length Practice Test: Crash Course

After finishing this book and taking our full-length practice exam, pick up REA's *Crash Course for AP Statistics*. Use your most recent score reports to identify any areas where you are still weak, and turn to the *Crash Course* for a rapid review presented in a concise outline style.

REA's Suggested 8-Week AP Study Plan

Depending on how much time you have until test day, you can expand or condense our eight-week study plan as you see fit.

To score your highest, use our suggested study plan and customize it to fit your schedule, targeting the areas where you need the most review.

	Review 1-2 hours	Quiz 15 minutes each	e-Flashcards Anytime, anywhere	Mini-Test 45 minutes	Full-Length Practice Test 3 hours
Weeks 1-2	Chapters 1-3	Quizzes 1 & 2	Access your e-flashcards from your computer or smartphone whenever you have a few extra minutes to study. Start with the 100 free cards included when you buy this book. Personalize your prep by creating your own cards for topics where you need extra study.		
Week 3	Chapter 4	Quizzes 3 & 4		Mini-Test 1 (The Mid-Term)	
Weeks 4-5	Chapter 5	Quizzes 5 & 6			
Weeks 6-7	Chapter 6	Quizzes 7 & 8		Mini-Test 2 (The Final)	
Week 8	Review Chapter 2 Strategies				Full-Length Practice Exam (Just like test day)

Need even more review? Pick up a copy of REA's *Crash Course for AP Statistics,* a rapid review presented in a concise outline style. Get more information about the *Crash Course* series at the REA Study Center.

Test-Day Checklist

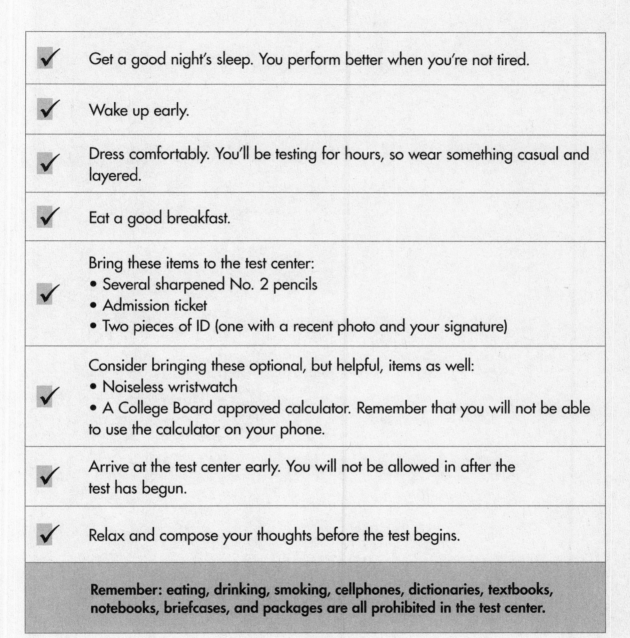

✓	Get a good night's sleep. You perform better when you're not tired.
✓	Wake up early.
✓	Dress comfortably. You'll be testing for hours, so wear something casual and layered.
✓	Eat a good breakfast.
✓	Bring these items to the test center: • Several sharpened No. 2 pencils • Admission ticket • Two pieces of ID (one with a recent photo and your signature)
✓	Consider bringing these optional, but helpful, items as well: • Noiseless wristwatch • A College Board approved calculator. Remember that you will not be able to use the calculator on your phone.
✓	Arrive at the test center early. You will not be allowed in after the test has begun.
✓	Relax and compose your thoughts before the test begins.

Remember: eating, drinking, smoking, cellphones, dictionaries, textbooks, notebooks, briefcases, and packages are all prohibited in the test center.

Strategies for the Exam

What Will I See on the AP Statistics Exam?

One May morning, you stroll confidently into the school library where you're scheduled to take the AP Statistics exam. You know your stuff: you paid attention in class, followed your textbook, took plenty of notes, and reviewed your coursework by reading a special test prep guide. You can identify and describe graphs, design experiments, and interpret test results. So, how will you show your knowledge on the test?

The Multiple-Choice Section

First off, you'll complete a lengthy multiple-choice section that tests your ability to not just remember facts about the various fields of statistics, but also to apply that knowledge to interpret and analyze statistical information. This section will require you to answer 40 multiple-choice questions in just 90 minutes. Here are the major fields of inquiry covered on the AP Statistics exam:

- Graphing and describing univariate, bivariate, and categorical data
- Experimental design and sampling methods
- Probability
- Simulations
- Inference procedures

So being able to name which type of sampling method was used if every tenth person was surveyed entering a basketball game (systematic, but you know that, right?) will not do you much good unless you can also explain how accurate sampling methods prevent bias and provide representative samples. It sounds like a lot, but by *working quickly and methodically* you'll have plenty of time to address this section effectively. We'll look at this in greater depth later in this chapter.

The Free-Response Section

After time is called on the multiple-choice section, you'll get a short break before diving into the free-response section. This section requires you to produce six written responses in 90 minutes. Like the multiple-choice section, the free-response portion of the exam expects you to be able to *apply your own knowledge to analyze data,* in addition to being able to provide essential facts and definitions.

What's the Score?

Although the scoring process for the AP exam may seem quite complex, it boils down to two simple components: your multiple-choice score plus your free-response scores. The multiple-choice section accounts for one-half of your overall score and is generated by awarding one point toward your "raw score" for each question you've answered correctly. The free-response section accounts for the remaining one-half of your total score. In this section, questions one through five count equally to make up three-eighths of your final score, and question six is weighted to count for one-eighth of your final score. Trained graders will read students' written responses and assign points according to grading rubrics. The number of points you accrue out of the total points possible for each question will form your score on the free-response section.

The test maker reports AP scores on a scale of 1 to 5. Although individual colleges and universities determine what credit or advanced placement, if any, is awarded to students at each score level, these are the assessments typically associated with each numeric score:

5 Extremely well qualified

4 Well qualified

3 Qualified

2 Possibly qualified

1 No recommendation

Section I: Strategies for the Multiple-Choice Section of the Exam

Because the AP exam is a standardized test, each version of the test from year to year must share many similarities in order to be fair. That means that you can always expect certain things to be true about your AP Statistics exam.

Which of the following phrases accurately describes a multiple-choice question on the AP Statistics exam?

(A) always has five choices

(B) may rely on a graph, table, or other visual stimulus

(C) may ask you to find a wrong idea or group related concepts

(D) more likely to test subject content than computation basics

(E) all of the above*

> Did you pick "all of the above?" Good job!

What does this mean for your study plan? You should focus more on the application and interpretation of the various analytical fields of statistics than on nuts and bolts such as formulas and finding P-values, because content makes up a much larger portion of the exam. Also, the required formulas will be provided to you on the test, and you will have access to your calculator. Keep in mind, too, that many statistical concepts overlap. This means that you should consider the connections among ideas and concepts as you study. This will help you prepare for more difficult interpretation questions and give you a head start on questions that ask you to use Roman numerals to organize ideas into categories. Not sure what this type of question might look like? Let's examine a typical Roman numeral item:

---

*Of course, on the actual AP Statistics exam, you won't see any choices featuring "all of the above" or "none of the above." Do, however, watch for "except" questions. We'll cover this kind of item a bit later in this section.

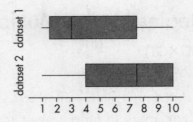

Which of the following statements accurately applies to the boxplots shown above?

 I. The range is the same for both graphs.

 II. Twenty-five percent of dataset 1 is equal to or greater than 50 percent of dataset 2.

 III. There are more data points in dataset 1.

 IV. The first quartile for dataset 1 is greater than the median for dataset 2.

(A) I and II

(B) I, III, and IV

(C) II, III, and IV

(D) I, II, and IV

(E) I, II, III, and IV

> Take a moment to look over the answer choices before evaluating each Roman numeral statement. You may notice that one numeral appears in more answer choices than do others. In this question, I and IV appear in the most choices. Evaluating those first can save you time; if IV is false, for example, the only possible correct answer is (A). Remember that the correct answer will include *all* of the applicable Roman numerals.

Types of Questions

You've already seen a list of the general content areas that you'll encounter on the AP Statistics exam. But how do those different areas translate into questions?

Question Type	Sample Question Stems
Graphs	*Which of the following would be greatest: mean, median, range, maximum, or IQR?*
Tables	*Find the conditional probability of A | B.*
Definition	*What type of sampling method was used?*

Question Type	Sample Question Stems
Example	*Which of the following would be considered a lurking variable?*
Factual	*Which of the following is NOT a reason to use blocking when designing an experiment?*
Interpretation of Results	*A p-value of 0.03 signifies ...*
Computer Output	*Which of the following represents the 95-percent confidence interval ...*

Throughout this book, you will find tips on the features and strategies you can use to answer different types of questions.

Achieving Multiple-Choice Success

It's true that you don't have a lot of time to finish this section of the AP exam. But it's also true that you don't need to get every question right to get a great score. Answering just two-thirds of the questions correctly—along with a good showing on the free-response section—can earn you a score of a 4 or 5. That means that not only do you not have to answer every question right; you don't even need to answer every question at all (although there is no penalty for wrong answers so you *should* answer every question). By *working quickly and methodically,* however, you'll have all the time you'll need. Plan to spend about two minutes and 15 seconds on each multiple-choice question. You may find it helpful to use a timer or stopwatch as you answer one question to help you get a handle on how long 135 seconds feels in a testing situation. Remember that you will be able to answer some questions very quickly while others will take longer. If timing is hard for you, set a timer for 33 minutes each time you take one of the online chapter-level quizzes that accompany this book, to help you practice working at speed. Let's look at some other strategies for answering multiple-choice items.

Process of Elimination

You've probably used the process-of-elimination strategy, intentionally or unintentionally, throughout your entire test-taking career. The process of elimination requires you to read each answer choice and consider whether it is the best response to the question given. Because the AP exam typically asks you to find the *best* answer rather than the *only* answer, it's almost always advantageous to read each answer choice. More than one choice may have some grain of truth to it, but one—the right answer—will be the most correct. Let's examine a multiple-choice question and use the process of elimination approach:

A group of researchers wants to find out if freshmen or seniors in high school are more flexible. They design a flexibility test that assigns to each student a flexibility score from 1 to 100. The researchers then administer the test to a random sample of freshmen and seniors attending the local high school. Which of the following would be an appropriate method to test the hypothesis that freshmen and seniors are equally flexible?

(A) 1-Proportion Z Test

(B) 2-Proportion Z Test

(C) 1-Sample T Test

(D) 2-Sample T Test

(E) Matched Pairs T Test

> To use the process of elimination, consider each option. Eliminate ideas that are clearly wrong, such as *1-Proportion Z Test* and *1-Sample T Test*, as we clearly have two sets of data. Cross out each choice as you eliminate it. Then consider the remaining choices. Which *method* reflects the type of data that has been gathered? If you're unsure, you can return to the question later or just guess. You've got a one-third chance of being right.

Students often find the most difficult question types on the AP exam to be those that ask you to find a statement that is *not* true or to identify an *exception* to a general rule. To answer these questions correctly, you must be sure to carefully read and consider each answer choice, keeping in mind that four of them will be correct and just one wrong. Sometimes, you can find the right answer by picking out the one that just does not fit with the other choices. If four answer choices relate to characteristics associated with quantitative data, for example, the correct answer choice may well be the one that relates to categorical data. Let's take a look at a multiple-choice question of this type.

The following can be said about a well-designed, completely randomized experiment EXCEPT:

(A) A cause-and-effect relationship can be determined.

(B) A control must be used.

(C) Replication is an essential component.

(D) A placebo is necessary.

(E) Randomization is used to assign the subjects or experimental units to treatment groups.

> To answer a NOT or EXCEPT question correctly, test each option by asking yourself: *Is this choice true? Does this correctly tell about a well-designed, completely randomized experiment?* A well-designed experiment includes randomization, replication, and control. It is the only way to show a cause-and-effect relationship. Which answer choice is not an essential feature of a well-designed experiment?

Predicting

Although using the process of elimination certainly helps you consider each answer choice thoroughly, testing each and every answer can be a slow process. To help answer the most questions in the limited time given AP test takers, you may find it helpful to instead try predicting the right answer *before* you read the answer choices. For example, you know that the answer to the math problem two-plus-two will always be four. If you saw this multiple-choice item on a math test, you wouldn't need to systematically test each response, but could go straight to the right answer. You can apply a similar technique to even complex items on the AP exam. Brainstorm your own answer to the question before reading the answer choices. Then, pick the answer choice closest to the one you brainstormed. Let's look at how this technique could work on a common type of question on the AP Statistics exam—one with computer output.

Students at one university are convinced that taking tests while hungry has adverse effects on test scores. The students take a census of their statistics class by asking their classmates to rate their hunger level on a scale from one (starving) to ten (full), and then having the professor match the hunger level with the test score results anonymously. The students wish to create a 95-percent confidence interval for the slope. The computer output is listed below:

Predictor	Coef	StDev	T	P
Constant	35.2547	1.45889	19.318	0.00378
Hunger Level	0.053649	0.375363	0.142927	0.886922
S = 7.00288		R-Sq = 0.0408%		R-Sq(adj) = 0.00211%

A computer output question will rarely ask you simply to read a piece of information from the output. Instead, you will need to apply what you have learned about statistics to interpret the information provided. You can use your knowledge of statistics to make good predictions to answer these types of questions. Always take a moment to study the computer output (or graph) before diving into the question.

In this example, notice that the output shows that the slope is 0.053649. It also shows that the standard error for the slope is 0.375363. So you already know that the answer is going to be $0.053649 \pm t^* (0.375363)$. You can then use the inverse t function on your calculator to find the correct answer: $0.053649 \pm 1.6759(0.375363)$. Now you can look at the answer choices and quickly pick your answer from the list.

A) 35.22547 ± 1.6759(1.45889)

B) 35.22547 ± 0.142927(0.375363)

C) 0.053649 ± 1.6759(1.45889)

D) 0.053649 ± 1.6759(0.375363)

E) 0.053649 ± 0.142927(0.375363)

What should you do if you don't see your prediction among the answer choices? Your prediction should have helped you narrow down the choices. You may wish to apply the process of elimination to the remaining options to further home in on the right answer. Then, you can use your statistical knowledge to make a good guess.

Learning to predict takes some practice. You're probably used to immediately reading all of the answer choices for a question, but in order to predict well, you usually need to avoid doing this. Remember, the test maker doesn't want to make the right answers too obvious, so the wrong answers are intended to sound like appealing choices. You may find it helpful to cover the answer choices to a question as you practice predicting. This will ensure that you don't sneak a peek at the choices too soon.

Sometimes, though, you need to have a rough idea of the answer choices in order to make a solid prediction, especially when there are lots of possible ways to interpret a question. Let's examine another question to practice predicting in this way.

When performing a hypothesis test, which of the following will give you the greatest power?

Before looking at the answer choices you would think to yourself: *To increase power I can increase sample size, increase* α*, or move* H_0 *further away from where I think the actual value is.* You know that the larger the sample size, the higher the power, but you are not going to arbitrarily start guessing sample sizes. You can now look at the answer choices and quickly see that (D) is the correct answer because it has the largest sample size and the largest alpha value.

A) $n = 200$, $\alpha = 0.01$

B) $n = 200$, $\alpha = 0.05$

C) $n = 600$, $\alpha = 0.01$

D) $n = 600$, $\alpha = 0.05$

E) Cannot be determined.

Avoiding Common Errors

Remember, answering questions correctly is always more important than answering every question. Take care to work at a pace that allows you to avoid these common mistakes:

- Missing key words that change the meaning of a question, such as *not, except,* or *least.* You might want to circle these words in your test booklet so you're tuned into them when answering the question.

- Overthinking an item and spending too much time agonizing over the correct response.

- Changing your answer but incompletely erasing your first choice.

Some More Advice

Let's quickly review what you've learned about answering multiple-choice questions effectively on the AP exam. Using these techniques on practice tests will help you become comfortable with them before diving into the real exam, so be sure to apply these ideas as you work through this book.

- Big ideas are more important than minutiae. Focus on learning important statistical concepts, models, and theories instead of memorizing formulas.

- You have just 135 seconds to complete each multiple-choice question. Pacing yourself during practice tests and exercises can help you get used to these time constraints.

- Because there is no guessing penalty, remember that making an educated guess is to your benefit. Remember to use the process of elimination to narrow your choices. You might just guess the correct answer and get another point!

- Instead of spending valuable time pondering narrow distinctions or questioning your first answer, trust yourself to make good guesses most of the time.

- Read the question and think of what your answer would be before reading the answer choices.

- Expect the unexpected. You will see questions that ask you to apply information in various ways, such as picking the wrong idea or interpreting a graph, table, or computer output.

Section II: Strategies for the Free-Response Section of the Exam

The AP Statistics exam always contains six free-response questions in its second section. This section always allows you 90 minutes to respond to all six of these questions. Often, these questions provide you with one or more visual stimuli, such as tables, computer output, and graphs. Then, the items ask a series of increasingly sophisticated questions. The question might begin by asking you to define one or two essential statistical concepts. Then, you may be asked to connect those definitions to the stimuli, or provide your own examples if no stimuli are provided. Finally, you may need to perform a sophisticated analysis of the statistical principle underlying the questions. This means that the free-response questions typically build in difficulty within themselves.

Students with a deeper understanding of the content tested in the item will normally receive higher scores on these items than students with a superficial knowledge of the content. Expect most of the free-response questions to require you to combine knowledge of different content areas—reading graphs and hypothesis tests, for example, or z-scores and confidence intervals—in order to fully respond.

The free-response section of the AP Statistics exam is unique. Unlike many other AP exams, such as history or English, you don't need to write a formal essay with an introduction and conclusion to answer the free-response questions on the AP Statistics exam. Nor is a list of numbers without any words an appropriate response. You'll need to write complete sentences that provide specific information requested in the various parts of a free-response question AND show all your mathematical calculations. Let's examine a typical free-response question.

Many cell phone plans allow families to share minutes. One hundred family plans were examined, each offering 1,400 minutes with unlimited texting. The following data were collected:

A text stimulus will precede the actual questions you must answer. This may provide additional information helpful in answering the questions.

Cell Phones with a Family Plan	Primary Phone (Account-holder)	Secondary Phone (Usually a second adult)	Third Phone (Usually oldest son or daughter)	Fourth Phone (Usually second oldest son or daughter)
n	100	100	92	57
Mean Minutes Used	156.2	273.0	199.8	245.1
Standard Deviation of Minutes Used	31.5	20.4	16.6	10.1
Mean Number of Text Messages Sent	41.3	97.2	506.1	382.4
Standard Deviation of Text Messages Sent	3.9	14.8	33.8	12.5

Many free-response questions provide a graph, table, computer output, or other visual stimulus. Study and interpret this as you construct your response.

a) If the distribution of minutes used is normal for each group of users (primary, secondary, third, and fourth), for whom would a bill that shows 215 minutes of use be most unusual?

Many free-response questions begin by asking you to define certain terms or perform simple calculations. Write your definitions in complete sentences, and show all of your work for any calculations.

b) Find a confidence interval for the number of minutes used each month by the primary user on the account.

> Questions parts will build in difficulty throughout the free-response item. Be sure to number or letter the parts of your written response to help essay scorers follow your thinking.

c) Using your confidence interval from part (b), test the hypothesis that the secondary user uses more minutes than the primary person on the billing statement.

> Questions parts will frequently refer to previous parts. Be sure to refer to part (b) when answering part (c). If for any reason you were unable to answer part (b), make up an answer and use that answer to write your solution to part (c). You may still receive full credit for part (c), although you will not receive any credit for part (b).

Achieving Free-Response Success

The single most important thing you can do to score well on the free-response section is to *answer the questions that you are asked*. It seems silly to point that out, doesn't it? But if you've ever written an essay or a even a research paper and received a mediocre grade because you didn't fully answer the question asked, or because you wrote about an almost-but-not-quite-right topic, you know how easy it can be to stray off topic or neglect to include all the facts needed in a written response. By answering each of the six free-response questions completely, you'll be well on your way to a great score on the AP exam. Let's look at some strategies to help you do just that.

Organizing Your Time

Although you have 90 minutes to write all six free-response items, you may choose to spend as long as you like on each individual question. The suggested time for free-response questions one through five is 13 minutes each. These questions are each worth 15 percent of your free-response score. The suggested time for the sixth question is 25 minutes; this question is worth 25 percent of your free-response score. Before you begin, take a few minutes to make a plan to address the section. *Scan* the first five questions and consider whether any of them seem especially difficult or easy to you. Then *read* question six. You can then plan to spend more time addressing the harder items, leaving less time for the simpler ones.

Let's discuss question number six, which is referred to as an investigative task. It will most likely ask you to perform a statistical operation that you have never encountered before. Do not let this intimidate you! There will be some parts of the question that you will know how to do. Make sure that you complete these parts of the question accurately. You will then be given instructions on how to complete a new statistical task or asked to complete a statistical calculation that is closely related to something that you have already learned. Don't worry that you have never done a problem exactly like this before; no one else taking the AP Statistics exam has either. Complete the question to the best of your ability and answer the questions in context. Trust that your statistical knowledge will carry over to this novel task.

Remember that you may also answer the questions in any order you wish. Start with the question about which you feel most confident. You want to make sure that you get all the points you can, and starting with a question that you are sure you can answer correctly is a great way to earn your first points. After completing one or two questions, move on to question number six. Remember, this investigative task is worth more toward your final score, so you do NOT want to leave it for last and run out of time. On the same note, you do not want to spend all your time on the investigative task question and neglect the other free-response questions. Watch your time, and after spending a reasonable amount of time on question number six, go back and answer another one of the earlier free-response questions. Budgeting your time is important. If you think one question will take you longer to answer, you can plan to spend less time addressing easier questions. Knowing the material for a given item especially well will probably mean it takes less time for you to write your response. Don't be concerned that you're not spending enough time on a given question if you know that you've written a good, thorough answer. You're being scored on content, not effort!

Let's look at another free-response question:

Pirates are famous for burying their treasure to keep it safe. Pirate Alfred makes lofty claims of being able to find all of his 1,213 burial locations. Pirate Bruno is much more realistic and admits he could relocate most of his loot, but not all 2,177 underground troves. Pirate Bruno, believing he can actually relocate more of his treasure than Pirate Alfred, decides to challenge Pirate Alfred to a contest to see which of the two pirates can relocate the treasure that they have buried from the last month. The pirates set off to uncover their wealth. Pirate Alfred uncovers 57 of his 89 caches from the last month. Pirate Bruno turns up 47 of his 58 hoards from the last month. If we assume that the buried treasure from the last month is a representative sample of all of the pirates' buried treasure, test Pirate Bruno's claim to dig up more of his fortune.

Think about what you will need to include to answer this question fully. You will want to ensure that you include in context the name of the test you will perform, the hypotheses, the conditions, the calculations, and your interpretation. To avoid forgetting any of these items, write down this list in your test booklet.

Stick to the Topic

Once you've noted what you will need to include, stick to it! As you write your response, you'll find that most of the hard work is already done, and you can focus on *expressing your ideas clearly, concisely, and completely.* Don't include your own opinions about the subject, and don't include extra information that doesn't help you fully answer the question asked and *only* the question asked. Essay scorers will not award you extra points for adding lots of irrelevant information or giving personal anecdotes.

Remember, too, that the essay scorers know what information has been provided in the stimulus. If a free-response question contains a graph, table, or computer output, don't waste time and effort describing the contents of the visual aids or the question content itself unless you are adding your own interpretation. In the question above, for example, writing *Pirate Bruno uncovered 47 out of 58 sites, and Pirate Alfred found 57 out of 89 sites* will not help your score any. If a question tells you that Pirate Bruno was a world-renowned pirate who was given the nickname "Robin Hood" for his tendency to steal from the rich and give to the poor, you do not need to restate Pirate Bruno's biography in your response or tell why you think Pirate Bruno was important—just calculate the proportion of his treasure he was able to find. Make your response short, to the point, and complete.

Make It Easy on the Scorers

As you're writing your responses, put yourself in the shoes of the AP Readers, who will read and consider your answers weeks from now. Expressing your ideas clearly and succinctly will help them best understand your point and ensure that you get the best possible score. Using your clearest handwriting will also do wonders for your overall score; free-response graders are used to reading poor handwriting, but that doesn't mean they can decipher every scribble you might make. Printing your answers instead of writing them in cursive may make them easier to read.

Another good way to point out your answers to scorers is to literally point them out with arrows and labels. Adding labels to each part of your response will help the AP Readers follow your response through the multiple parts of a free-response question, and can only help your score.

Revision

Even the best statisticians make mistakes, especially when writing quickly: skipping or repeating words, forgetting to interpret a result, or neglecting to write an answer in context are all common errors that occur when someone is rushed. Reserving a few minutes at the end of your free-response time will allow you to review quickly your responses and make necessary corrections. Ensuring that you have interpreted your results and written your answers in context (this includes labels for all of your answers) are the two most important edits you can make to your writing; these are necessary to formulate complete responses for your calculations and will help your score.

Remember that free-response graders are not mind readers, so they will only grade what's on the page, not what you thought you were writing. At the same time, remember, too, that graders will not grade anything you have crossed out. So you don't need to spend time erasing errors; simply cross it out and it will not be graded.

A Sample Response

After you have read, considered, planned, written, and revised, what do you have? A thoughtful free-response written answer likely to earn you a good score, that's what. Remember that free-response graders must grade consistently in order for the test to be fair. That means that all AP Readers look for the same ideas in each answer to the same question using a rubric. Let's examine how a reader might apply that technique to grading a response.

Pirates are famous for burying their treasure to keep it safe. Pirate Alfred makes lofty claims of being able to find all of his 1,213 burial locations. Pirate Bruno is much more realistic and admits he could relocate most of his loot, but not all 2,177 underground troves. Pirate Bruno, believing he can actually relocate more of his treasure than Pirate Alfred, decides to challenge Pirate Alfred to a contest to see which of the two pirates can relocate the treasure that they have buried from the last month. The pirates set off to uncover their wealth. Pirate Alfred uncovers 57 of his 89 caches from the last month. Pirate Bruno turns up 47 of his 58 hoards from the last month. If we assume that the buried treasure from the last month is a representative sample of all of the pirates' buried treasure, test Pirate Bruno's claim to dig up more of his fortune.

Two-Proportion Z-Test

$H_0: P_A = P_B$

$H_a: P_A < P_B$

Where P_A is the proportion of treasure Pirate Alfred could relocate, and P_B is the proportion of treasure Pirate Bruno could relocate.

Conditions:

SRS (problem states to assume this)

$n_A p_A > 10$ $57 > 10$

$n_A q_A > 10$ $32 > 10$

$n_B p_B > 10$ $47 > 10$

$n_B q_B > 10$ $11 > 10$

Population $> 10n$ Pirate Alfred: $1{,}213 > 10 \times 89 = 890$

                    Pirate Bruno: $2{,}177 > 10 \times 58 = 580$

We will assume the two samples are independent and that the pirates are not sabotaging each other's efforts.

Calculations:

2-Proportion Z Test

$Z = -2.21309$

p-value $= 0.013446$

$\hat{P}_A = 0.640449$

$\hat{P}_B = 0.810345$

Pooled $\hat{P} = 0.707483$

$n_A = 89$

$n_B = 58$

Conclusion:

We reject H_0 because our p-value is low ($0.013 < 0.05$). There is evidence that Pirate Bruno can relocate more of his treasure than Pirate Alfred. If the null hypothesis was true, we would observe differences this large 1.3 percent of the time due to chance.

Anytime you are performing an inference test, always start by naming the test you will perform and stating the null and alternative hypotheses.

Don't forget to state clearly in your hypothesis which group is which.

Don't leave out your conditions!

For two sample inference procedures, check your conditions for both groups.

Your calculator has done all of the calculations for you. Make sure you write down all of the information given. It is better to write down something extra than to omit something that was needed.

Your conclusion should be stated in context of the problem. If the problem is about pirates, talk about pirates!

Here, an interpretation of the p-value is also given. It was not required for this problem, but if you get in the habit of always including the interpretation of the p-value, you will have it in your answer when you need it.

Some More Advice

What have you learned about the free-response section? Keep these ideas in mind as you prepare for the AP Statistics exam. Becoming comfortable with these techniques will make you feel confident and prepared when you take the exam in May.

- You don't have to spend the suggested time on each question. One question may be very straightforward and you may not need as much time to answer it. If you know it, write your response and move on.

- Consider previewing all the free-response questions before you start to write. You can choose to answer the questions in the manner that best suits you. Either answer the hardest questions first and leave the easiest for last when you're tired, or start with the questions you are sure you will be able to earn the most points on.

- Show all of your work! If you used your calculator to find an answer, write down what you typed in and what the answer was, but don't write down calculator notation.

- Stay on topic and answer the question! Addressing the question fully is the single most important way to earn points on this section.

- Answer in context. Statistics is largely about interpreting your results. Writing a p-value without connecting it to the context of the question will earn you very little credit.

- Label everything. Put titles, labels, and scales on your graphs. Label your answers with the correct units. A single number by itself is rarely the correct answer.

- If you make a graph on your calculator, sketch it on your paper. The AP Readers can't see what was on your calculator during the test unless you write it down on your paper.

- Handwriting is important and must be legible! If the AP Reader can't read your writing, you'll get no points, even if your response is correct.

- Be sure to label the parts of your responses. Make it easy for the AP Reader to award you points by being able to easily navigate your response. If you make it clear and easy for the reader, you'll earn the reward!

- Leave a few minutes to quickly review and revise your answers. You don't need to double-check all of your answers, but you do need to make sure that you have answered each question completely and in context. Leave space as you write in case you need to add something later (perhaps you forgot a condition) and to make it easier for the scorer to read your handwriting.

Two Final Words: Don't Panic!

The free-response questions can and probably will ask you about concepts and data you haven't thought about before. The possibilities are practically endless. Remember that all free-response questions seek to test your knowledge of statistical theories and concepts, not highly specific facts. Applying what you know to these unfamiliar concepts and data will help you get a great score, even if you've never thought much about the particular data listed in the question.

Exploring Data

Constructing and Interpreting Graphical Displays of Distributions of Univariate Data

Variables

When data are collected from a group of individuals, the characteristics under study are called **variables**. The variables could be heights in feet of 15 mountains in North America, the number of cars owned by a family, or the colors of silk blouses on sale in 12 stores in Los Angeles.

Variables can be either **categorical** or **quantitative** (numerical). Categorical variables are those whose values take on names or labels, such as the colors of blouses. Quantitative variables are those that show quantity and come from counting or measuring.

Quantitative variables can be further classified as **discrete** or **continuous**. Discrete numerical variables usually come from counting and can take on only certain values, such as the number of cars owned by a family. A family can have 2 cars or 3 cars, but cannot have 2.5 cars. The height of a mountain is an example of a continuous variable—one from measuring. It could take on practically any value, such as 1000 feet, 1001 feet, or 1000.437 feet.

Data are often classified by the number of variables under study. **Univariate** data are data where only one variable is being considered at a time. This variable may be from a single group or from multiple groups being compared. When two variables are being studied simultaneously, the data are called **bivariate**. Bivariate data are covered later in this chapter.

Describing Distributions of Quantitative Data

Data can be organized by listing or plotting the **observed values** of the variable in a **frequency table** or graph. Below is an example of a frequency table.

	Scores of AP Statistics Students				
Score	1	2	3	4	5
Frequency	10	18	38	28	17

Graphical displays include the **bar chart**, **dotplot**, **stemplot**, **box-and-whisker plot**, **histogram**, and **cumulative frequency graph**. Each of the graphs listed will be discussed in detail momentarily. (Bar charts are used for categorical data and are covered in Section E.)

Center, shape, and spread are characteristics that describe the distribution of univariate data. The center can be estimated by locating the middle of the distribution. Estimate the point where about half of the data are on either side of that point. You may have learned that the **mean** and **median** are measures of center. They will be discussed in more detail in Section B. For now, to find the center of a distribution from a graph, estimate as described above.

To describe the shape of the distribution, we use the terms such as symmetric, bell- or mound-shaped, skewed, uniform, unimodal, and bimodal. When examining the shape of the data, we also look for gaps, outliers, or any other unusual features.

Symmetric graphs appear to have mirror images about their center. If a graph has only one clear peak, it is called **unimodal**; if it has two, it is **bimodal**. Symmetric, unimodal graphs may sometimes be referred to as **mound-shaped**, or **bell-shaped**, because they look like a mound or bell.

Skewed graphs are unimodal graphs that tend to slant—most of the data are clustered on one side of the distribution and "tails" off on the other side. If the tail is on the left, we call the distribution left-skewed. If the tail is on the right, it is right-skewed.

A **uniform** distribution is symmetric where the data are distributed fairly evenly across the graph. There are no clear peaks and the data do not seem to cluster in one area or another.

Below are some examples of graphs and their shapes.

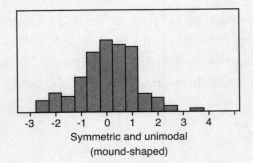

Symmetric and unimodal
(mound-shaped)

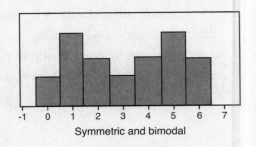

Symmetric and bimodal

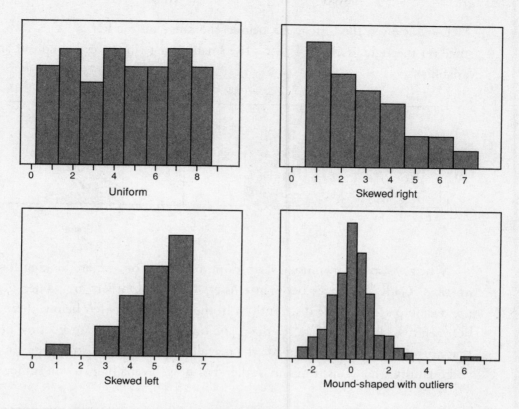

The spread describes the **variability** of the distribution of the data.

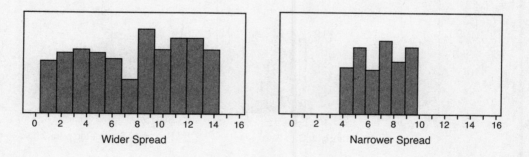

The histogram on the left is more variable than the one on the right. Note that they are graphed on the same scale and it is clear which graph has more spread.

When interpreting the spread of a graphical display consider the **range**—the **maximum** minus the **minimum** value—as a starting point. If there seem to be values outside of the general pattern, you would do well to also look at how variable the data are without them. You may have learned that the **standard deviation** and **interquartile range** are also used to measure the variability of the data. These will be discussed in more detail in Section B. For now, look at the overall spread of the data, both with and without any unusual observations.

Look again at the histograms below. The range on the left is $14 - 1 = 13$ while the range on the right is $10 - 4 = 6$. The smaller range indicates less spread and, thus, less variability.

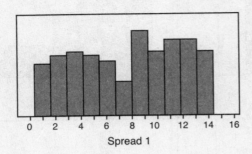
Spread 1

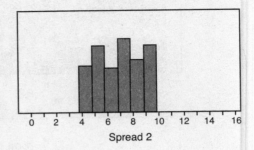
Spread 2

When we speak of unusual features of a distribution, we are looking for **gaps** and/or **outliers**. Gaps are spaces between clusters of data. Outliers are values that fall out of the overall pattern of the distribution. In the graph on the left below, there is a clear gap between two clusters of data. In the graph on the right, there are a couple of observations that are outliers—they fall outside the overall pattern of the distribution. In Section B we will quantify what constitutes an outlier. For now, "eyeballing" it will suffice.

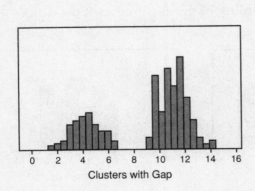
Clusters with Gap

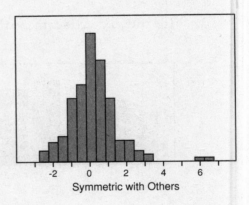
Symmetric with Others

Remember, when describing distributions think *center*, *spread*, *shape*, and *outliers/ unusual features*.

Constructing Displays of Quantitative Data

A **dotplot** is a quick method to plot small sets of data, usually discrete, that are not too spread out. Computer software can make dotplots of large sets of data, but doing so by hand takes time. The dotplot below represents the number of people living in a house for 15 residences on a particular block in a small town.

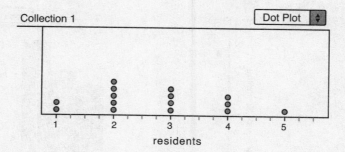

Each dot represents one house, and the number of dots in each column represents how many houses have a certain number of residents. There are two houses with one resident in the house, five houses with two residents, and so on.

EXAMPLE: Describe the distribution below of the number of people living in the 15 residences.

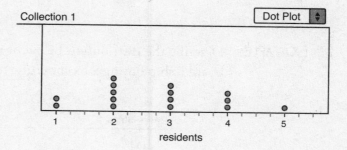

ANSWER: The distribution has a shape slightly skewed to the right with a peak at two residents. The data seem to be centered at three residents and the range of the data is four residents. There are no extreme values present or gaps within the data.

TEST TIP

Beginning in May 2011, the AP Exam stopped penalizing test takers for incorrect responses to multiple-choice questions. Entering a response for every question—even a wild guess— may help improve your score.

A **histogram** is an appropriate display for quantitative data. It is used primarily for continuous data, but may be used for discrete data that have a wide spread. The horizontal axis is broken into intervals or **bins**. Histograms are also good for large data sets. The histogram below shows the amount of money spent by passengers on a board ship during a recent cruise to Alaska.

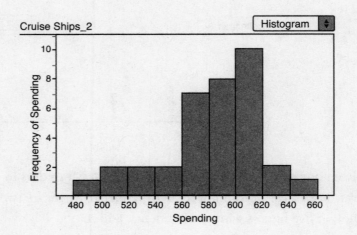

The bins have widths of $20. One person spent between $480 and $500, two spent between $500 and $520, and so on. We cannot tell from the graph the precise amounts each individual spent.

EXAMPLE: Describe the distribution below of the amount spent by passengers on board a ship during a recent cruise to Alaska.

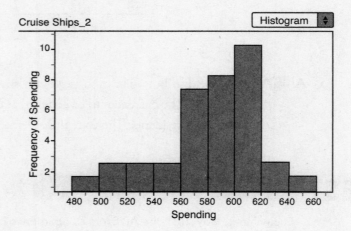

ANSWER: The distribution has a shape skewed to the left with a peak around $600 to $620. The data are centered at about $590—this is about where half of the observations will be to the left and half to the right. The range of the data is about $180, but the clear majority of passengers spent between $560 and $620. There are no extreme values present or gaps within the data.

DIDYOUKNOW?

An estimated 19.2 million people worldwide went on a cruise in 2011, about 4 percent more than had done so the previous year.

The vertical axis on the previous histogram is of count or frequency. It can also be a relative frequency or percentage. This is often useful if a data set is very large. Below is an example of a relative frequency histogram.

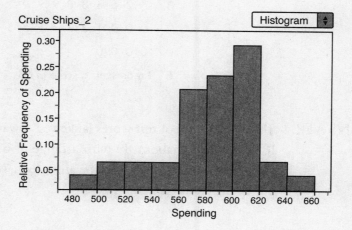

A **stemplot**, also called stem-and-leaf plot, can be used to display univariate data as well. It is good for small sets of data (about 50 or less) and forms a plot much like a histogram. The stemplot below represents test scores for a class of 32 students.

```
   3 | 3
   4 |
   5 |
   6 | 3 7 9
   7 | 2 2 5 7
   8 | 1 2 6 8 8 8 9 9 9
   9 | 0 0 0 1 3 3 4 5 5 6 7
  10 | 0 0 0 0
Key: 6 | 3 represents a score of 63
```

The values on the left of the vertical bar are called the stems; those on the right are called leaves. Stems and leaves need not be tens and ones—they may be hundreds and tens, ones and tenths, and so on. A good stemplot always includes a key for the reader so that the values may be interpreted correctly.

◾ **EXAMPLE:** Describe the distribution of test scores for students in the class using the stemplot below.

```
 3 | 3
 4 |
 5 |
 6 | 3 7 9
 7 | 2 2 5 7
 8 | 1 2 6 8 8 8 9 9 9
 9 | 0 0 0 1 3 3 4 5 5 6 7
10 | 0 0 0 0
```

Key: 6 | 3 represents a score of 63

◾ **ANSWER:** The distribution of test scores is skewed toward lower values (to the left). It is centered at about 89 with a range of 67. There is an extreme low value at 33, which appears to be an outlier. Without it, the range is only 37, about half as much.

Sometimes, the shape of a stemplot is hard to describe because one has only a few stems but many leaves. Consider the plot below.

```
1 | 0 0 1 4 4 5 6 6 8 8 9
2 | 2 2 3 3 7 7 8
3 | 0 0 1 1 1 3 3 4 5 5 5 6 6 7 7 7 7 9
4 | 0 0 0 0 1 1 1 2 2 3 3 4 5 6 6 6 6 8 8 8 8 9 9 9
```

A split-stem plot is the solution. "Split" the stems by having two stems of "1," two of "2," and so on. Leaves with values 0–4 would be placed in the first of the split stems and values 5–9 would be placed in the second of the split stems.

```
1 | 0 0 1 4 4
1 | 5 6 6 8 8 9
2 | 2 2 3 3
2 | 7 7 8
3 | 0 0 1 1 1 3 3 4
3 | 5 5 5 6 6 7 7 7 7 9
4 | 0 0 0 0 1 1 1 2 2 3 3 4
4 | 5 6 6 6 8 8 8 8 9 9 9
```

If we did not split the stems, it would be more difficult to view the shape of the distribution.

Calculator Tip:

Many graphs and functions for univariate data can be done on the graphing calculator, but the observations must be first entered into a list. We will use the test scores of the 32 students from a previous example: 33, 63, 67, 69, 72, 72, 75, 77, 81, 82, 86, 88, 88, 88, 89, 89, 89, 90, 90, 90, 91, 93, 93, 94, 95, 95, 96, 97, 100, 100, 100, 100.

From the home screen, press STAT.

```
EDIT CALC TESTS
1:Edit...
2:SortA(
3:SortD(
4:ClrList
5:SetUpEditor
```

Choose 1:Edit. This brings you into the main list editor.

```
L1      L2      L3      1
━━━━━   ──────  ──────



L1(1)=
```

Enter the data into the list that you are going to use. For now, use L1. Press ENTER after each entry to move to the next line.

```
L1      L2      L3      1
33      ──────  ──────
63
67
69
72
72
━━━━━
L1(7)=
```

Continue to type in values until the list is complete. Note that the cursor is on the last entry, which in this case is 100, that is the 32nd entry into this list.

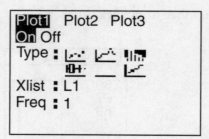

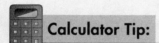
Calculator Tip:

Histograms can be made from data in lists. From the home screen, press $\boxed{2^{nd}}$ $\boxed{\text{STAT PLOT}}$, and then choose 1:Plot1....

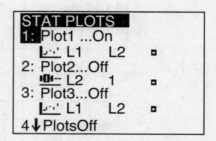

Turn on the plot by pressing $\boxed{\text{ENTER}}$ when the cursor is over On. Arrow right to the third type of plot, which is a histogram. Press $\boxed{\text{ENTER}}$ so that it is highlighted. Arrow down to Xlist: and press $\boxed{2^{nd}}$ $\boxed{\text{L1}}$. On the Freq: line, enter 1, if it is not already so. (This is the frequency of each value in Xlist and is used if two lists act as a frequency table.)

Press ZOOM 9 to see the graph. The calculator will sometimes choose bin widths that are hard to use.

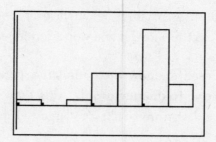

Press WINDOW . A bin width of 5 or 10 may make more sense in this situation than the one of 11.16. Change Xmin, Xmax, Xscl, Ymin, Ymax, and Yscl to the values shown below on the right. Xscl is the width of your bins, beginning at Xmin. An appropriate window for these data is shown on the screen below.

```
WINDOW
 Xmin=33
 Xmax=111.16666...
 Xscl=11.166666...
 Ymin=-4.20966
 Ymax=16.38
 Yscl=.1
 Xres=1
```

```
WINDOW
 Xmin=30
 Xmax=105
 Xscl=5
 Ymin=-4
 Ymax=15
 Yscl=1
 Xres=1
```

Press GRAPH to view the histogram.

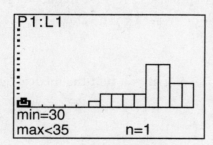

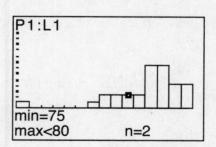

You can use the TRACE button and arrows to view the frequencies of each bin.

Note: Before making a plot, it is good practice to make sure that any function entered into the Y= editor is turned off or deleted so that it does not interfere with the plots that will be graphed. You can turn off a function by pressing Y= ; arrow to the function you want turned off by placing the cursor over "=" next to that function, and then press ENTER to deselect it. Now, it will not interfere with your plot.

The display used to show the cumulative percentage of values in a distribution is a **cumulative relative frequency graph**. This type of plot shows the sums of the relative frequencies of the data from smallest to largest. The display below shows the ages of people who took a river cruise.

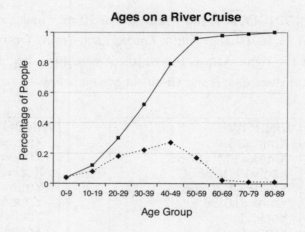

The dotted line represents the percentage of passengers in each age group; the solid line represents the cumulative percentage up to and including that age group. The last age group (80–89) has a cumulative percentage of 100, which means that all of the people on the cruise were aged 89 and below. A cumulative relative frequency graph helps us locate what percentage of a distribution is at or below a given value.

EXAMPLE: Using the river cruise plot above, find the median age of people on this cruise.

ANSWER: The median is the value where 50% of the observations are at or below it. Find the 50% (0.5) mark on the vertical axis, and move to the right from that point until you intersect the solid line. Move down from that point to the horizontal axis, which shows the age group on the cruise. In this case the median is in the age group from 30 to 39. We cannot be sure exactly where it is in that interval.

Summarizing Distributions of Univariate Data

Measuring Center: Mean and Median

The mean and median are **measures of center**. They locate the middle or the center of a distribution.

The **mean** is the average of a data set. The mean is calculated by the formula, $\bar{x} = \dfrac{\sum x_i}{n}$, where n is the number of observations in the data set, and x_i represents individual observations counting from 1 to n. The summation symbol, is the command to add up the expression following it.

EXAMPLE: The weights (in pounds) of 10 jockeys that are going to race at Arlington Park are 113, 117, 112, 113.5, 115.8, 114, 114.6, 113.5, 112.4, and 113. Calculate the mean weight of the jockeys.

ANSWER:
$$\bar{x} = \frac{\sum x_i}{n}$$
$$= \frac{113 + 117 + 112 + 113.5 + 115.8 + 114 + 114.6 + 113.5 + 112.4 + 113}{10}$$
$$= \frac{1138.8}{10}$$
$$= 113.88 \text{ pounds.}$$

The **median** is the middle number of a data set that has been arranged from the smallest to largest value or vice versa. If there is an even number of values in the data set, the median is the mean of the two in the middle.

EXAMPLE: What is the median of the data set 4, 5, 5, 6, 6, 7, 7, 7, 9?

ANSWER: The median is 6 since it is the middle value.

EXAMPLE: What are the mean and median of the set 4, 5, 5, 6, 6, 7, 7, 7, 9, 10?

ANSWER: The median is 6.5 since the mean of the middle two numbers, 6 and 7, is 6.5. Note that the median does not have to be one of the data points.

The mean is the sum of the data divided by the number of data, or $\frac{66}{10} = 6.6$.

The median is a better measure of center if the data set has outliers or is skewed in any way. The maximum value in the previous example was 10. If it were larger, the median would not be affected, but the mean would, since the mean is computed using the actual values of the data.

EXAMPLE: What are the mean and median of the set 4, 5, 5, 6, 6, 7, 7, 7, 9, 30?

ANSWER: The median is still 6.5, as it was in the previous example. The median is not affected by extreme values or outliers. The mean for the data set is 8.6.

The shape of a distribution can tell us about the relationship between the mean and median. In symmetric distributions, the mean and median are close to the same. In skewed distributions, the mean is "pulled" away from the median toward the tail. Since the mean is based on the values of the data, it tends to go toward more extreme values, which in skewed distributions are located in the tail.

EXAMPLE: Estimate the median of each distribution shown below and describe how the mean compares to it.

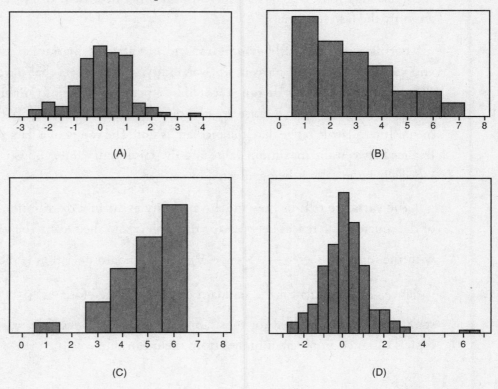

(A)

(B)

(C)

(D)

ANSWER: Graph (a) is symmetric with a median at about 0. The mean will also be about 0. Graph (b) is right-skewed with a median of about 3. The mean will be pulled more toward the values of 4–7 than it will be toward 1–2, so the mean will be greater than 3. Graph (c) is left-skewed with a median of 5. The mean will be pulled toward the lower values, so it is less than 5. Graph (d) is symmetric with a few extreme high values. The median is about 0, and the mean will be just slightly larger because of the outliers.

Measuring Spread: Range, Interquartile Range, Standard Deviation, Variance, and Outliers

In addition to measures of center, distributions need to be described with measures of **variability**. It is not enough to know where the middle of a distribution is, but also how spread out it is. A manufacturer of light bulbs would like small variability in the amount of hours the bulbs will likely burn. A coach who needs to decide which athletes go on to the finals may want larger variability in heat times because it will be easier to decide who are truly the fastest runners.

For data that are fairly **symmetric** and not affected by outliers, **standard deviation** and **variance** are useful measures of variability. For data sets that have extreme values or skewness, the **interquartile range** would be a better measure of variability. This is because standard deviation and variance are computed with the actual values of the data, like the mean. Interquartile range, like the median, is not. The **range** of a data set is simply the difference between the maximum value and the minimum value, and is rarely a good choice, especially because of it being affected by outliers.

The **variance** tells us how much variability exists in a distribution. It is the "average" of the squared differences between the data values and the mean. The variance is calculated with the formula $s^2 = \dfrac{1}{n-1}\sum(x_i - \overline{x})^2$. The standard deviation is the square root of the variance. The formula for the standard deviation is therefore $s = \sqrt{\dfrac{1}{n-1}\sum(x_i - \overline{x})^2}$. The standard deviation is used for most applications in statistics. It can be thought of as the typical distance an observation lies from the mean.

EXAMPLE: The average monthly rainfall in inches in Birmingham, England, is shown in the table below.

Jan	Feb	Mar	Apr	May	Jun	Jul	Aug	Sep	Oct	Nov	Dec
2.3	1.9	2.1	1.8	2.2	2.2	2.0	2.8	2.2	2.1	2.5	2.6

Compute the variance and standard deviation of the monthly rainfall.

ANSWER: To compute the variance and standard deviation, the mean must first be computed. In this data set, $\overline{x} = \dfrac{26.7}{12} = 2.225$. The variance is computed as follows:

$$
\begin{aligned}
s^2 &= \frac{1}{n-1}\sum(x_i - \overline{x})^2 \\[2mm]
&= \frac{1}{12-1}(2.3-2.225)^2 + (1.9-2.225)^2 + (2.1-2.225)^2 + \cdots + (2.6-2.225)^2 \\[2mm]
&= \frac{1}{11}(0.9225) \\[2mm]
&= 0.084.
\end{aligned}
$$

The standard deviation is the square root of the variance, or $\sqrt{0.084} \approx 0.290$. The monthly mean rainfall in Birmingham is 2.225 inches and it varies by about 0.29 inches from that each month—sometimes more, sometimes less.

The **five-number summary** is composed of the minimum, maximum, median, **first quartile** (Q_1), and **third quartile** (Q_3) of a data set. The quartiles and the median (which is the second quartile or Q_2) break the data into four equally sized groups. These groups may not have the same spreads, though.

To find Q_1, determine the middle value of all observations below the median in an ordered data set. The value of Q_3 is found at the middle value of all observations above the median of a data set.

The **interquartile range** (***IQR***) is a measure of variability that works well for data that are skewed or have outliers. The *IQR* is the spread of the middle 50% of the data and not affected by extreme values. The *IQR* is calculated by $Q_3 - Q_1$. The interquartile range is a single value, like the range.

EXAMPLE: Find the five-number summary and interquartile range of the data set 4, 5, 5, 6, 6, 7, 7, 7, 9, 30.

ANSWER: The minimum value is 4, the maximum is 30, and the median is 6.5. The first quartile is the middle of the five values below the median, or 5. The third quartile is the middle of the five values above the median, or 7. The five-number summary is min = 4, Q_1 = 5, median = 6.5, Q_3 = 7, max = 30.

The interquartile range is $Q_3 - Q_1 = 7 - 5 = 2$

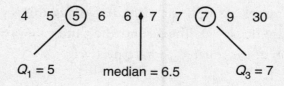

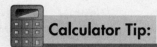

Summary statistics can be computed using the graphing calculator. We will compute these values for the rainfall data from Birmingham, England.

Jan	Feb	Mar	Apr	May	Jun	Jul	Aug	Sep	Oct	Nov	Dec
2.3	1.9	2.1	1.8	2.2	2.2	2.0	2.8	2.2	2.1	2.5	2.6

From the home screen, press STAT . Choose 1:Edit. Enter the data into L1.

```
L1        L2        L3      1
2
2.
2.2
2.1
2.5
2.6

L1(13) =
```

Press 2nd QUIT to return to the home screen.

Press STAT , arrow right to CALC, and then choose 1:Var Stats. Press 2nd L₁ and then press ENTER .

```
EDIT CALC TESTS
1: 1-Var Stats
2: 2-Var Stats
3: Med-Med
4: LinReg(ax+b)
5: QuadReg
6: CubicReg
7↓ QuartReg
```

```
1-Var Stats L1
```

The first entry on the screen, $\bar{x}$, is the mean; the fourth entry Sx is the standard deviation of the sample. (x is the standard deviation if this is an entire population.) The sample size, n, is also displayed. The second and third values, x and x^2, are the sum of the data and the sum of the squared data, respectively.

Arrowing down gives the five-number summary.

Outliers are extreme values that do not fall within the general pattern of the data. A common rule for determining whether an observation is an outlier is the $1.5IQR$ rule.

If an observation is farther than $1.5IQR$ above the third quartile, or farther than $1.5IQR$ below the first quartile, it is considered an outlier. Those boundaries at $Q_1 - 1.5IQR$ and $Q_3 + 1.5IQR$ are called "fences." Outliers are points lying outside the fences.

EXAMPLE: The salaries of the New York Yankees (in thousands of dollars) can be summarized by the five-number summary:

Minimum	Q_1	Median	Q_3	Maximum
323	2,375	5,667	12,678	26,000

Would the maximum or minimum salaries be considered outliers?

ANSWER: The interquartile range of the salaries is $12{,}678 - 2{,}375 = 10{,}303$. The lower outlier fence is located at

$$Q_1 - 1.5IQR = 2{,}375 - 1.5(10{,}303) = -13{,}709.5.$$

The upper outlier fence is located at

$$Q_3 + 1.5IQR = 12{,}678 + 1.5(10{,}303) = 28{,}132.5.$$

Neither the minimum nor the maximum values are beyond the fences, and are therefore not considered outliers.

DIDYOUKNOW?

In 2011, the New York Yankees had the highest combined payroll in Major League Baseball. Pitcher C. C. Sabathia brought in the team's highest salary at just over $24 million.

Constructing Box-and-Whisker Plots

A **box-and-whisker plot**, or more simply a **boxplot**, is a graphical display of the five-number summary and any outliers. The "box" of the boxplot is drawn from the first quartile to the third quartile and then is split into two sections by the median. The box shows the spread of the middle half of the data. Whiskers are usually extended from the ends of the box to the minimum and maximum values. If outliers are present, as determined by the $1.5IQR$ rule, the whiskers only extend to the most extreme values inside the fences. Outliers are then plotted as points by themselves.

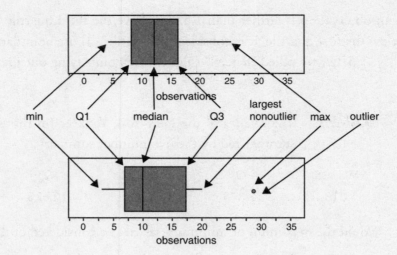

■ **EXAMPLE:** The salaries of the New York Yankees (in thousands of dollars) can be summarized by the five-number summary:

Minimum	Q_1	Median	Q_3	Maximum
323	2,375	5,667	12,678	26,000

Construct a box-and-whisker plot for the distribution and describe its shape.

■ **ANSWER:** We saw in the previous example that there are no outliers. So, we merely need to plot these five values to scale.

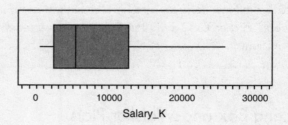

The shape of the distribution is skewed right. Each quarter of the data, moving from low salaries to high salaries, has a wider spread than the previous one. The data cluster toward the left and spread out toward the right—a right skew.

A graphing calculator boxplot and histogram are shown below to emphasize this point.

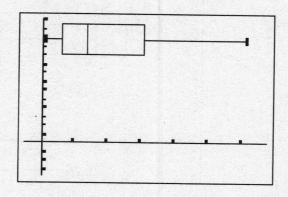

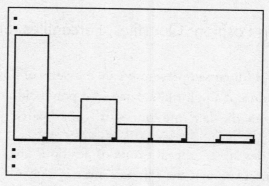

![Calculator Tip icon] **Calculator Tip:**

Boxplots can be created on the graphing calculator. The Birmingham weather data have been entered into L1.

Jan	Feb	Mar	Apr	May	Jun	Jul	Aug	Sep	Oct	Nov	Dec
2.3	1.9	2.1	1.8	2.2	2.2	2.0	2.8	2.2	2.1	2.5	2.6

From the home screen, press 2nd STAT PLOT and then choose 1:Plot1.... Highlight the boxplot icon, leftmost in the bottom row. Complete the rest of the entries as shown. Mark: is the symbol that will show outliers, if any exist.

Press ZOOM 9 to see the graph in an appropriate viewing window.

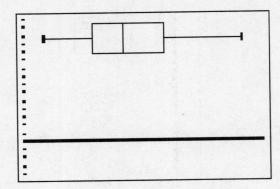

Measuring Position: Quartiles, Percentiles, Standardized scores

Quartiles (discussed previously) are measures of **position**. They give information relating the location of a point in a data set or population to the rest of the group. Recall that quartiles divide the data into four equally sized parts.

Percentiles are another measure of position. They divide the data set into 100 equal parts. An observation at the Pth percentile is higher than P percent of all observations. If you took a standardized test and your score was in the 83rd percentile, 83% of the people that took the same test scored lower than you. If you scored in the 83rd percentile on another standardized test, you were still in the same position relative to everyone else on *that* test. The size of the population that took the test may have changed, but that is not indicated in the percentile score.

Standardized scores, or **z-scores**, are a way to measure where an individual in a population stands relative to the mean by using the standard deviation as a unit of measure. z-scores measure how many standard deviations away from the mean (in either direction) an observation is.

A z-score is calculated by the formula $z = \dfrac{\text{observation} - \text{mean}}{\text{standard deviaton}}$. z-scores that are positive represent observations that are greater than the mean; negative z-scores represent observations that are less than the mean.

EXAMPLE: You scored an 87 on a test in your Statistics class where the mean was 85 and the standard deviation was 3. Your best friend is in a different Statistics class and scored a 90 where the mean in her class was 88 and the standard deviation was 4. Who had the better score relative to their own class?

ANSWER: For you, $z = \dfrac{87 - 85}{3} = 0.667$. For your friend, $z = \dfrac{90 - 88}{4} = 0.5$. You scored better relative to your class since your z-score is larger. The z-scores represent how many standard deviations from the mean each of your scores were.

Computer Printout of Summary Statistics

You will sometimes be given a computer printout summarizing a set of data. The output is fairly easy to read, but there are some things of which you should be aware. Below is a sample computer printout.

N	Mean	Median	TrMean	StDev	SEMean
80	4.58	4.04	4.37	2.99	0.33

Min	Max	Q1	Q3
0.46	13.63	2.15	6.56

The sample size (N), mean, median, standard deviation (StDev), mininum, maximum, first quartile, and third quartile are easily identified. The two values we have not covered in this section are TrMean and SEMean.

TrMean is the "trimmed mean." It is usually the mean of the middle 90% of the observations—potential outliers have been "trimmed" off. SEMean is the "standard error of the mean." It will be covered in more detail in Chapters 3 and 4.

The Effect of Changing Units on Summary Measures

Changing units will change measures of center and spread by the same ratio as the multiplier.

EXAMPLE: The weights (in pounds) of 10 jockeys that are going to race at Arlington Park are 113, 117, 112, 113.5, 115.8, 114, 114.6, 113.5, 112.4, and 113. Summary statistics are as follows: mean = 113.88, standard deviation = 1.54, median = 113.5, IQR = 1.6.

If the jockey's weights were converted from pounds to kilograms (2.2 pounds = 1 kilogram), what would the new summary measures be?

ANSWER: Converting each observation from pounds to kilograms (dividing by 2.2), we get

Pounds	113	117	112	113.5	115.8	114	114.6	113.5	112.4	113
Kilograms	51.4	53.2	50.9	51.6	52.6	51.8	52.1	51.6	51.1	51.3

The new summary measures are as follows:

	Pounds	Kilograms
Mean	113.88	51.76
Standard deviation	1.54	0.70
Median	113.5	51.6
IQR	1.6	0.73

Each of the summary measures in kilograms is the equivalent measure in pounds divided by 2.2. It is not necessary to convert the individual observations. One only need convert the summary measures.

Adding or subtracting the same constant to or from each will change measures of center in a similar manner, but will not change measures of spread.

EXAMPLE: Each jockey's clothes and saddle add about 4 pounds to the weight placed on the horse. What are the summary statistics for the weight of the 10 jockeys plus their equipment?

ANSWER: One need not add 4 pounds to each jockey's weight and recompute the summary statistics. Measures of center will increase by 4 pounds, and measures of spread will not change. The summary measures are as follows:

	Jockey	Jockey plus equipment
Mean	113.88	117.88
Standard deviation	1.54	1.54
Median	113.5	117.5
IQR	1.6	1.6

TEST TIP

Be sure to bring at least two pencils and a sturdy eraser to the AP Exam. Traditional pencils are better than mechanical pencils because the tips are less likely to break. Using a slightly dull pencil can also help you fill in bubbles more quickly, shaving valuable seconds off the time it takes to answer each multiple-choice question.

Comparing Distributions of Univariate Data (Dotplots, Back-to-Back Stemplots, Parallel Boxplots)

When comparing two or more sets of univariate data it is important to look at similarities and differences between the groups as well as any interesting features within the groups. When making comparisons, be clear about your statements. If you mention that one distribution has a center of 16 and the other distribution has a center of 20, a direct comparison of that relationship should be stated. In other words, state that the center is four units larger for the second distribution than the first. This should also be followed for measures of spread, measures of location, and the shapes of the distributions.

You must also consider that both distributions may be compared to some benchmark, like whether weight was lost or gained—the benchmark is no change—or whether a passing score was attained on an exam. Sometimes this benchmark may be subtly implied; sometimes it will be explicitly stated.

The keys to remember: *center, spread, shape,* and *outliers/unusual features.*

EXAMPLE: The weight loss of 30 people for 1 week on Diet Plan X is shown in the first dotplot and the weight loss of 30 people for 1 week on Diet Plan Y is shown in the second dotplot. A positive value represents a loss in weight; a negative value represents a gain in weight. Compare the two diet plans.

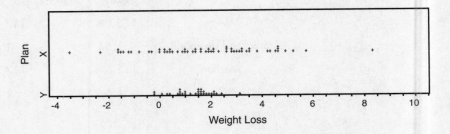

ANSWER: Both distributions are fairly symmetric with Diet Plan X having an average weight loss of around 1.8 pounds while Diet Plan Y has an average weight loss of around 1.4 pounds. People on Plan X lost an average of 0.4 pounds more than Plan Y.

There is more variability in Plan X than in Plan Y. The overall spread of Plan X is about three times that of Plan Y, and Plan X has one extremely high value. Even though the average loss for Plan Y is less than Plan X, Plan Y only had 2 people actually gain weight whereas Plan X had 10 people gain weight, and Plan Y's gains were much smaller. More people lost larger amounts of weights with Plan X than with Plan Y.

Note: You may be asked on the exam which diet plan you would consider using for weight loss. It would be important to mention the fact that even though Plan Y had only a few people that lost small amounts of weights as compared to Plan X, you still might consider Plan X since many people lost larger amounts of weights. No matter which you choose, you must complete your reasoning with reference to *both* diet plans, not just the one you are choosing.

EXAMPLE: Below are parallel boxplots representing pretest and posttest scores for a written driving learner's permit test where 8 represents the score needed to receive a learner's permit. Students took a weekend class on driver safety before taking the posttest. Compare the two distributions.

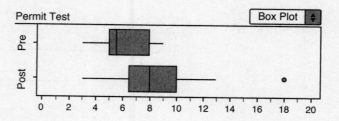

ANSWER: Both distributions are fairly symmetric overall, but the middle half of the pretest scores may have some skew. The posttest scores contain a high outlier at 18. The average on the posttest was about 2.5 points higher than the pretest—8 vs. 5.5. The posttest scores are more variable than the pretest scores, having nearly double the spread, not counting the outlier. Only one-fourth of the test takers scored well enough on the pretest to get a permit whereas half of the test takers on the posttest scored well enough to get a permit. It appears that the training course improved scores on the test overall.

Calculator Tip:

The graphing calculator can display parallel boxplots. Two sets of data have been entered into the following lists:

L1: 13, 15, 15, 15, 16, 18, 19, 20, 20, 22, 25, 28, 30, 31, 31, 35, 38, 39, 40, 43

L2: 28, 30, 33, 34, 36, 36, 37, 39, 40, 43, 43, 44, 44, 46, 47, 48, 48, 49, 52, 53

From the home screen, press 2nd STAT PLOT . Set up Plot1 and Plot2 for boxplots where Plot1 will display the data in L1 and Plot2 will display the data in L2.

```
STAT PLOTS
1: Plot1...On
    ΗΠ⊢-- L1    1      □
2: Plot2...On
    ΗΠ⊢-- L2    1      □
3: Plot3...Off
    ⌐··⌐ L1    L2     □
4↓ Plots Off
```

Press ZOOM 9 to view both plots on the same scale of an appropriate viewing window.

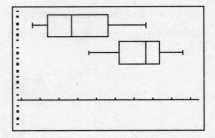

Exploring Bivariate Data

Analyzing Patterns in Scatterplots

Bivariate data consist of two variables, between which one is typically looking for an association. The variables may be categorical or quantitative; in this section we will focus on quantitative bivariate data.

The two variables under study are referred to as the **explanatory variable** (x) and the **response variable** (y). The explanatory variable *explains* or *predicts* the response variable. The response variable measures the outcomes that have been observed.

EXAMPLE: Data collected from snack foods included the number of grams of fat per serving and the total number of calories in the food. Identify the explanatory and response variables when looking for a relationship between fat grams and calories.

ANSWER: The explanatory variable is grams of fat and the response variable is calories. The number of grams of fat would be a predictor of the number of calories in the snack.

Scatterplots are used to visualize quantitative bivariate data. These plots can tell us if and how two variables are related. When examining univariate data we described a distribution's shape, center, spread, and outliers/unusual features. In a scatterplot, we will focus on its shape, direction, and strength, and look for outliers and unusual features. Below is a scatterplot of the top 30 leading scorers in the National Basketball Association (NBA). Each point represents 1 of the 30 players. Michael Jordan, who scored 32,292 points in 1,072 games, is noted.

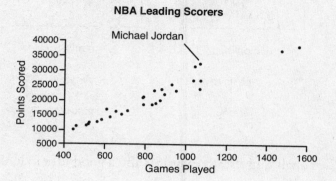

The **shape** of a plot is usually classified as linear or nonlinear (curved). The **direction** of a scatterplot tells what happens to the response variable as the explanatory variable increases. This is the slope of the general pattern of the data. The **strength** describes how tight or spread out the points of a scatterplot are.

The table below shows comparisons of scatterplots of various shapes, directions, and strengths.

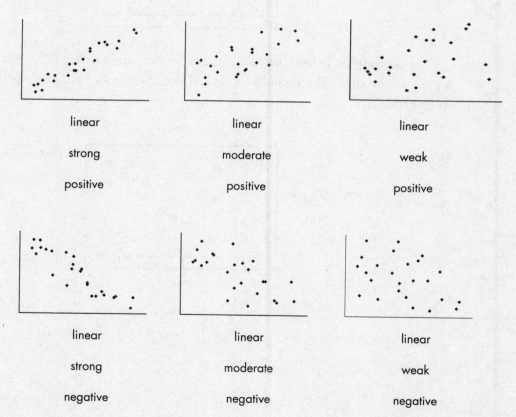

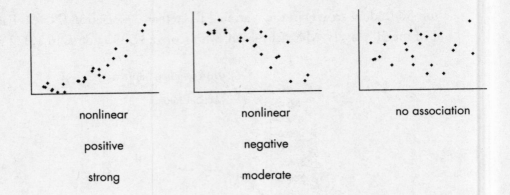

nonlinear	nonlinear	no association
positive	negative	
strong	moderate	

When analyzing a scatterplot it is also a good idea to look for outliers, clusters, or gaps in the data. The scatterplot below has an obvious gap. There is an overall positive, linear association but we should find out the reason for the gap.

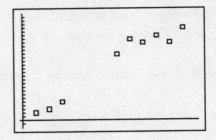

The scatterplot below has an obvious outlier. An outlier falls outside the general pattern of the data. There could be several possible reasons for the outlier and it merits investigation.

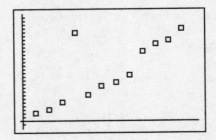

EXAMPLE: A scatterplot of the top 30 scorers in NBA history is shown below. Identify the explanatory variable and the response variable. Describe the association between the two variables.

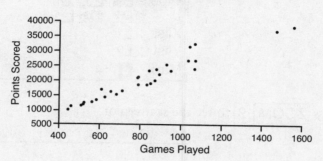

NBA Leading Scorers

ANSWER: The explanatory variable is the number of games played. The response variable is the number of points scored. The relationship is linear, strong, and positive. There are no outliers, but there is a large gap between 1,100 and 1,400 games. The two extreme points, perhaps a couple of players with very long careers, appear to follow the general pattern of the data.

Calculator Tip:

A scatterplot can be viewed on the graphing calculator. First, the data must be entered into lists. Recall that you access the list editor by pressing $\boxed{\text{STAT}}$ and then choosing 1:Edit....

L1	L2	L3	2
2	1	------	
3	3		
5	7		
6	7		
9	11		
10	12		

L2(7) =

Press $\boxed{2^{\text{nd}}}$ $\boxed{\text{STAT PLOT}}$ and choose 1:Plot 1. Turn on the plot, select the scatterplot icon, and enter the appropriate lists for Xlist: and Ylist:.

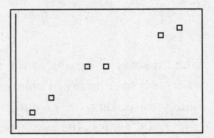

Press $\boxed{\text{ZOOM}}$ $\boxed{9}$ to see the scatterplot.

Correlation

The strength of a *linear* association can be measured by the **correlation coefficient**. This notation for correlation is r and is calculated using the following formula:

$$r = \frac{1}{n-1} \sum \left(\frac{x_i - \bar{x}}{s_x} \right) \left(\frac{y_i - \bar{y}}{s_y} \right).$$

The correlation coefficient is based on the means and standard deviations of the two variables. It is the average product of the standardized scores of x and y for each point. There are several important characteristics of the correlation coefficient.

- Correlation can only be calculated with quantitative variables.

- The value of r has a maximum of 1 and a minimum of -1. When r is exactly 1 or -1, all of the data points fit exactly on a line. A correlation of 0 indicates no linear association. The sign of the correlation indicates the direction of the relationship.

- Correlation does not change when the variables are interchanged, i.e., when the explanatory becomes the response and the response becomes the explanatory.

- Since r is calculated using standardized values, the correlation value will not change if the units of measure are changed (changing pounds to kilograms, for example).

- Correlation only measures strengths of *linear* relationships and is not an appropriate measure for nonlinear data.

- Correlation is not a resistant measure. It is affected by outliers.

Correlation is a unitless number that has no direct interpretation. However, some benchmarks can help us get a feel for the strength of a linear association. Ignoring the sign, correlations less than 0.5 are considered weak, those between 0.5 and 0.8 are considered moderate, and those greater than 0.8 are considered strong. These are not absolute rules, but do give us a frame of reference.

EXAMPLE: Match each scatterplot below with the correlation that best describes the linear association between the explanatory and response variables.

$$r = 0.468 \qquad r = 0.998 \qquad r = -0.503$$
$$r = 0.069 \qquad r = -0.940$$

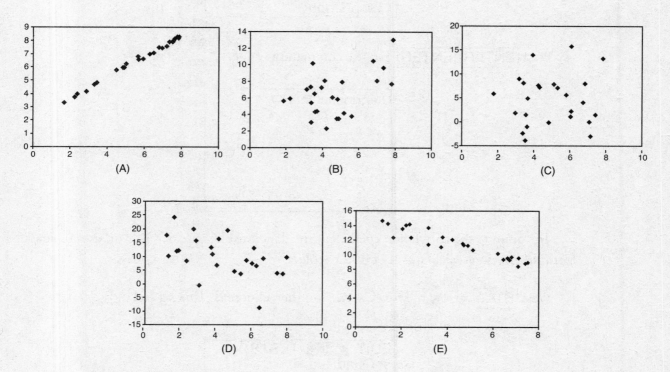

ANSWER: A. 0.998   B. 0.468   C. 0.069   D. 20.503   E. 20.94

Graph A has the strongest correlation; graph C has the weakest correlation. Graphs B and D have about the same strength; it is only the direction that is different.

Calculator Tip:

The graphing calculator can compute the correlation coefficient.

When using the TI family of graphing calculators, it is important to make sure that the diagnostics feature is turned on in order to view the correlation. You only need to do this once. (You will have to do it again if the calculator is reset or the batteries are changed.)

Press $\boxed{2^{nd}}$ $\boxed{\text{Catalog}}$ and arrow down to $\boxed{\text{DiagnosticOn}}$.

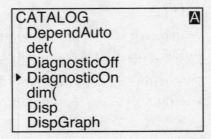

Press $\boxed{\text{ENTER}}$ $\boxed{\text{ENTER}}$ and the command is done.

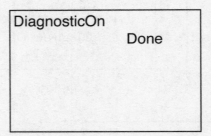

To compute the correlation coefficient the data must be in lists. (This tip uses the data from the previous calculator tip on scatterplots.)

Press $\boxed{\text{STAT}}$, arrow right to CALC, and then choose 8: LinReg (a1bx).

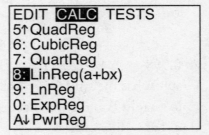

Enter the list containing x (in this case L1), a comma, and then the list holding y (in this case L2). Press ENTER .

```
LinReg(a+bx) L1,
L2
```

The correlation coefficient is the value of r.

```
LinReg
  y=a+bx
  a=−.9508196721
  b=1.33442623
  r²=.9750654855
  r=.9874540422
```

A strong correlation does *not* mean that there is a cause–effect relationship between the two variables. You need an experiment to justify a cause–effect relationship. There may be other variables that contribute to the relationship seen. One such variable is a **lurking variable**. This is a variable that has a simultaneous effect on both the explanatory and response variables, but was not part of the study. It may cause changes in both variables under study and therefore be the reason for the association.

Least-Squares Regression Line

Linear regression is a method of finding the best model for a linear relationship between the explanatory and response variables. This method calculates the best-fit line by minimizing the sum of the squares of the differences between the observed values and the predicted values from the line. This line is called the **least-squares regression line** or LSRL.

The LSRL has the equation $\hat{y} = b_0 + b_1 x$, where b_0 is the y-intercept of the line and b_1 is the slope. The symbol $\hat{y}$ is read "y-hat" and is the *predicted* value of the response variable y for the explanatory variable x.

The slope and intercept are calculated by the following formulas: $b_1 = r\dfrac{s_y}{s_x}$ and $b_0 = \bar{y} - b_1\bar{x}$, where $\bar{x}$ and $\bar{y}$ are the means of the explanatory and response variables, respectively; s_x and s_y are the respective standard deviations.

Notation alert! The formulas provided on the AP Statistics Exam use b_0 for the intercept and b_1 for the slope. Graphing calculators and some textbooks use a for intercept and b for slope. You must be able to use these interchangeably.

The least-squares regression line is an *average*. It predicts the mean value of y for a given value of x. This property also means that the least-squares regression line *always* goes through the point $(\overline{x}, \overline{y})$.

EXAMPLE: The heights (in inches) and shoe sizes for 15 women were recorded. The equation of the least-squares regression line is $\hat{y} = -17.7 + 0.390x$, where x is the height and y is the shoe size. What is the predicted shoe size for a woman 63 inches tall?

ANSWER: Substitute 63 for height (x) into the equation. The predicted shoe size is $\hat{y} = -17.7 + 0.390x = -17.7 + 0.390(63) = 6.87$. Since this is not an actual shoe size, we might conclude that a woman 63 inches tall would wear a shoe of size 6.5 or 7. However, the least-squares regression model predicts the *average* shoe size for 63-inch-tall women to be 6.87.

We would not want to use this model to predict the shoe size of a woman 78 inches tall. Seventy-eight inches is outside the range of the data that were used to fit the model. Using the model to predict that shoe size is called **extrapolation**. Responses predicted in this way could be unreasonable. Shoe sizes may not necessarily increase at the same rate for people over 6 feet tall and under 5 feet tall. Be careful of extrapolation in any model.

The slope of the least-squares regression line is also an average. It describes the mean change in the predicted value of the response variable for a one-unit change in the explanatory variable.

EXAMPLE: The heights (in inches) and shoe sizes for 15 women were recorded. The equation of the least-squares regression line is $\hat{y} = -17.7 + 0.390x$, where x is the height and y is the shoe size. Interpret the slope of the LSRL.

ANSWER: The slope is 0.390, which means that for every increase of 1 inch in women's heights, their shoe sizes increase by an average of 0.39.

DID YOU KNOW?

Approximately 68 percent of U.S. women have shoe sizes between 6.5 and 9.5. The same percentage of U.S. women measure between 62 and 67 inches tall.

The **coefficient of determination** or r^2 is another way of indicating how well the linear model fits the data. It is computed by squaring the correlation r. Unlike correlation, though, it does have an interpretation. The value of r^2 is the percent of variation in the response variable that can be explained by its linear relationship with the explanatory variable.

EXAMPLE: A scatterplot of the heights (in inches) and shoe sizes for 15 women is shown below, along with the least-squares regression line. The value of r^2 is 0.851. Interpret r^2.

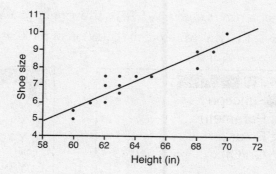

ANSWER: These is variability in the shoe sizes and there is variability in the heights. In this case, about 85.1% of the variability in the shoe size can be explained by its linear relationship with the height. The other 14.9% is due to other variables not under consideration.

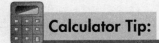

Calculator Tip:

The least-squares regression model can be computed with the graphing calculator. This tip uses the previous data lists as shown below.

L1	L2	L3	2
2	1	------	
3	3		
5	7		
6	7		
9	11		
10	12		

L2(7) =

Press, $\boxed{\text{STAT}}$ arrow right to CALC, and then choose 8: LinReg (a1bx). Enter the list containing x (in this case L1), a comma, and then the list holding y (in this case L2). *Do not* press $\boxed{\text{ENTER}}$ yet.

LinReg(a+bx) L1,
L2, ∎

Now, enter a comma, press $\boxed{\text{VARS}}$, arrow right to Y-VARS, and choose 1:Function…. Then, choose 1:Y1. This will store the equation of the LSRL in Y1.

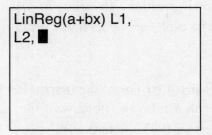

VARS **Y-VARS**
1: Function...
2: Parametric...
3: Polar...
4: On/Off...

FUNCTION
1: Y1
2: Y2
3: Y3
4: Y4
5: Y5
6: Y6
7↓Y7

LinReg(a+bx) L1,
L2,Y1

Press ENTER now. The intercept, slope, coefficient of determination, and correlation are displayed.

```
LinReg
  y=a+bx
  a=−.9508196721
  b=1.33442623
  r²=.9750654855
  r=.9874540422
```

If you press ZOOM 9, you will see the scatterplot and the least-squares regression line overlaid.

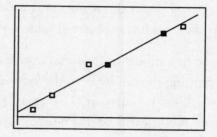

Computer Printout of Least-Squares Regression

On the exam, you may be provided with computer output to interpret. The output below is of the women's height and shoe size data.

The regression equation is

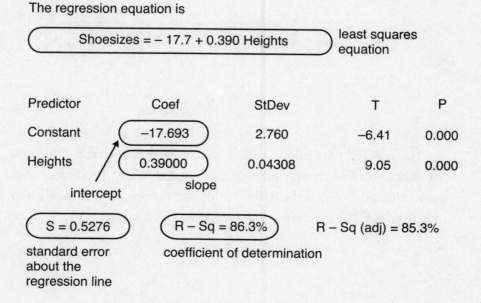

Shoesizes = − 17.7 + 0.390 Heights   least squares equation

Predictor	Coef	StDev	T	P
Constant	−17.693	2.760	−6.41	0.000
Heights	0.39000	0.04308	9.05	0.000

intercept   slope

S = 0.5276   R − Sq = 86.3%   R − Sq (adj) = 85.3%

standard error
about the
regression line   coefficient of determination

Note that the correlation coefficient r does not show up on the printout. It can be found by taking the square root of R-Sq(uared): $r = \sqrt{0.863} = 0.929$. Do not forget that the sign of the correlation is the same as the slope, which is positive in this case. Also, it would be best to ignore R-Sq(adj), as its use is not part of the AP Statistics curriculum.

Finally, the value of s is the standard error about the regression line. Like standard deviation is the typical distance a value lies from the mean, the value of s here is the typical vertical distance a point is from the LSRL. If you were to make a prediction, it would give you an estimate of about how much you might over- or underpredict.

Residual Plots, Outliers, and Influential Points

Residuals are the differences between the observed values of the response variable y and the predicted values $\hat{y}$ from the regression model. There is one residual for each point, which is calculated as residual = observed value − predicted value = $y - \hat{y}$.

After you have fit a model to the data, it is best to examine the graph of the residuals. This is done by plotting the residuals on the vertical axis against the explanatory variable. If the residual graph shows no curve or "U-shaped" pattern, a linear model is appropriate for the data. However, if a linear model is not appropriate, the residual graph will have some sort of pattern or curved feature.

The graph below is the residual plot for the height and shoe size data. There is no curved pattern in the plot, so a linear model is appropriate for these data.

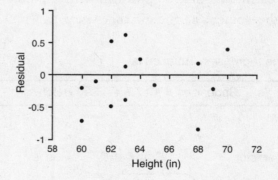

The residual plot below is from a data set that is clearly not linear. Even though the correlation coefficient may be high, implying that any linear association would be strong, the residuals tell us that the pattern of the data is not truly linear. This graph gives us reason to believe that the data are not well modeled by the line.

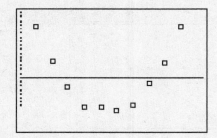

A residual plot can also reveal possible outliers. **Outliers** have large residuals since those points are far from the least-squares regression line (in the *y*-direction). An outlier usually does not have a great effect on the slope of a least-squares regression line, but does have influence on the correlation.

The scatterplots below represent a data set with an outlier (left) and with the outlier removed (right). The least-squares regression is shown for both sets. Note that the slope (*b*) did not significantly change when removing the outlier, but the correlation (*r*) increased substantially.

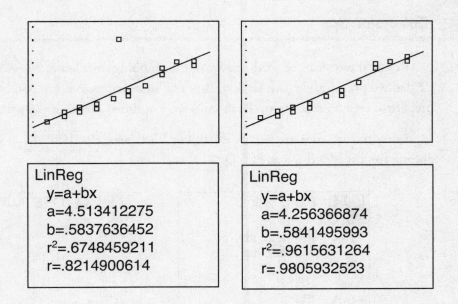

Influential points are extreme points that radically affect the slope of a LSRL. For example, consider the scatterplots below of men's heights and shoe sizes with their regression lines and values of r^2. The graph on the left shows data for 15 randomly selected men. The graph on the right shows the same men plus Shaquille O'Neal, an 85-inch tall basketball player with a shoe of size 22.

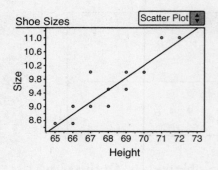

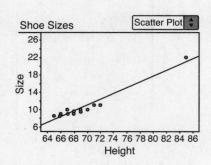

If O'Neal's point were removed from the data, the slope would drop from $0.676 \frac{\text{sizes}}{\text{inch}}$ to $0.357 \frac{\text{sizes}}{\text{inch}}$. Therefore, O'Neal is an influential observation. Note also that the removal of O'Neal would reduce r^2 from 0.94 to 0.79. This is also a characteristic of extreme points that seem to be in the general pattern of the data—they artificially strengthen correlation and r^2.

 Calculator Tip:

Residual plots can be made easily on the graphing calculator. We will use the lists L1 and L2 defined previously in this section. Run the linear regression function LinReg (a1bx); then prepare to create a scatterplot. Both tasks were outlined earlier in this section.

Enter the appropriate list for Xlist:. For Ylist: press $\boxed{2^{\text{nd}}}$ $\boxed{\text{LIST}}$, arrow down, and then choose the list identified at RESID.

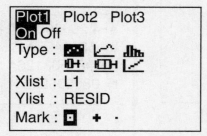

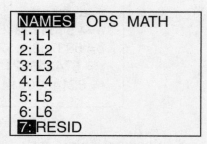

Press $\boxed{\text{ZOOM}}$ $\boxed{9}$ to see the residual plot.

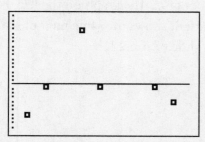

TEST TIP

Transformations to Achieve Linearity: Logarithmic and Power Transformations

When the regression calculations and residual plots indicate that a linear model is not appropriate, we must find another model for the data. This involves transforming the response variable and in some cases the explanatory variable, as well. Remember that regression and correlation are used when describing and analyzing linear data only. When we transform data to achieve linearity, we can use the regression techniques as defined in the previous section. At that point, predictions can be made and the prediction can be transformed back to terms of the original data.

A common method of transforming data is a **logarithmic transformation**. This is done by taking the logarithm of a variable and using it in place of the original variable when creating a scatterplot and finding the least-squares regression model. Two models that can be obtained by logarithmic transformations are the **exponential model** $\hat{y} = ab^x$ and the **power model** $\hat{y} = ax^b$. For an exponential model, only the response variable is transformed; for the power model, both are transformed.

A certain bacteria's population has been recorded as follows over time (hours) as shown below. The original data show a curved pattern. Populations can often be modeled by an exponential function of the form $\hat{y} = ab^x$.

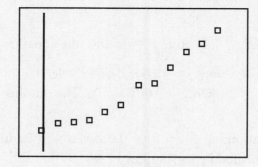

The logarithm of the y-values was taken. A scatterplot of $(\log y)$ versus x is shown below along with its least-squares regression line.

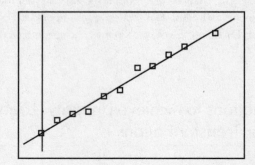

The graph of the residuals from the transformed linear regression shows that a linear model is appropriate for the transformed data.

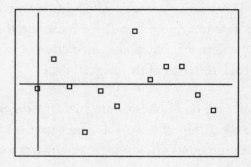

Since a linear model is acceptable for the transformed data, the original data fit an exponential model. You can now use the transformed linear model to make predictions.

EXAMPLE: The linear model of $\log y$ on x is given by the equation $\hat{y} = 1.85 + 0.065x$ where x is the time in hours and y is the population of bacteria in thousands. Predict the number of bacteria at time $x = 16$ hours.

ANSWER: Substitute $x = 16$ into the equation

$\log \hat{y} = 1.85 + 0.065x = 1.85 + 0.065(16) = 2.89$. Since $\log \hat{y} = 2.89$, then $y = 10^{2.89} \approx 776$. The number of bacteria is about 776 thousand.

The equation $\log \hat{y} = 1.85 + 0.065x$ is written iterms of $\log \hat{y}$ and not $\hat{y}$. It can be **back-transformed** by solving for $\hat{y}$.

$$\log y = 1.85 + 0.065x$$
$$10^{\log y} = 10^{1.85+0.065x}$$
$$y = 10^{1.85} \cdot 10^{0.065x}$$
$$y = 70.8(1.161)^x.$$

Predictions using this exponential function will yield the same values as with the transformed linear model. That is, $\hat{y} = 70.8(1.161)^x = 70.8(1.161)^{16} \approx 771$. (The difference is due to the round-off error.)

We also use logarithmic transformations for power models. A power model has a curved look to it just as the exponential model. The power model is written as $\hat{y} = ax^b$.

The scatterplot below shows the relationship between x and y.

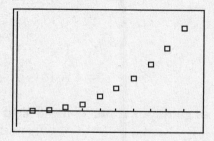

Assume that simply doing a logarithmic transformation of y did not straighten the plot. We may take the logarithm of *both* the explanatory and response variables. A scatterplot of $(\log y)$ versus $(\log x)$ is shown below along with its residual plot.

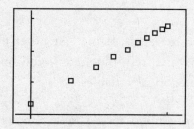 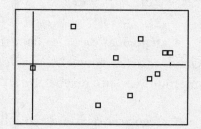

Since there is no U-shaped pattern in the residual graph, a power model was appropriate. You can now use the transformed linear model to make predictions.

EXAMPLE: The linear model of log y on log x is given by the equation log $\log \hat{y} = -0.356 + 1.484(\log x)$, where x is a planet's distance from the Sun (in millions of miles) and y is its orbital period (in years). Predict the orbital period of Mars, a planet 141 million miles from the Sun.

ANSWER: Substitute $x = 141$ into the equation.

$\log \hat{y} = -0.356 + 1.484(\log 141) \approx 2.83$. Since, $\log \hat{y} = 2.83$ then $y \approx 681$. The orbital period of Mars is about 681 days.

The equation $\log \hat{y} = -0.356 + 1.484(\log x)$ is written in terms of log $\hat{y}$ and log x, not y and x. It can be back-transformed by solving for $\hat{y}$.

$$
\begin{aligned}
\log y &= -0.356 + 1.484(\log x) \\
10^{\log y} &= 10^{0.356 + 1.484(\log x)} \\
y &= 10^{0.356} \cdot 10^{1.484(\log x)} \\
y &= 0.441 x^{1.484}.
\end{aligned}
$$

Predictions using this power function will yield the same values as with the transformed linear model. That is, $\hat{y} = 0.441 x^{1.484} = 2.270(141)^{1.484} \approx 682$. (The difference is due to the round-off error.)

Exploring Categorical Data

Categorical data are data that are labels, names, or other nonnumerical outcomes. Gender (male, female), car color (white, red, blue, etc.), and one's response to a survey question (yes, no, maybe) are all categorical variables.

Categorical data are typically organized into a table, known as a **frequency table**. A frequency table summarizes how many times a particular category of a variable occurred.

EXAMPLE: A survey of 12 adults asked, "Of chocolate, strawberry, and vanilla, which is your favorite ice cream flavor?" The results were as follows:

Flavor

vanilla	strawberry	vanilla
chocolate	vanilla	strawberry
vanilla	chocolate	chocolate
vanilla	chocolate	vanilla

Create a frequency table that summarizes these data.

ANSWER:

Flavor	Chocolate	Strawberry	Vanilla
Frequency	4	2	6

Note: A frequency table of one variable, such as the one above, is sometimes called a **one-way table**.

The quantity of each response can be expressed as a **relative frequency**, which is a percentage or proportion of the whole number of data. A table of relative frequencies is known as a **relative frequency table**.

EXAMPLE: A survey of 12 adults asked, "Of chocolate, strawberry, and vanilla, which is your favorite ice cream flavor?" The results were as follows:

Flavor

vanilla	strawberry	vanilla
chocolate	vanilla	strawberry
vanilla	chocolate	chocolate
vanilla	chocolate	vanilla

Create a relative frequency table that summarizes these data. Round relative frequencies to three decimal places.

ANSWER: The relative frequencies of chocolate, strawberry, and vanilla are $\frac{4}{12}, \frac{2}{12}$, and $\frac{6}{12}$ respectively.

Flavor	Chocolate	Strawberry	Vanilla
Frequency	0.333	0.167	0.500

OR

Flavor	Chocolate	Strawberry	Vanilla
Frequency	33.3%	16.7%	50.0%

As with numerical data, categorical data can be represented visually in graphs or charts. One method is a **bar chart**. A bar chart displays the distribution of a categorical variable, with each bar's length proportional to the frequency of its represented category.

■ EXAMPLE: Create a bar chart showing the frequencies of responses as given in the following frequency table.

Flavor	Chocolate	Strawberry	Vanilla
Frequency	4	2	6

■ ANSWER:

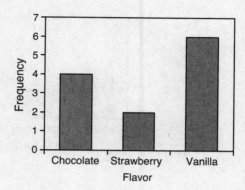

A bar graph can display relative frequencies as well.

EXAMPLE: Create a bar chart showing the relative frequencies of responses as given in the following relative frequency table.

Flavor	Chocolate	Strawberry	Vanilla
Relative Frequency	33.3%	16.7%	50.0%

ANSWER:

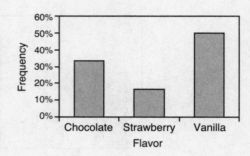

Two-Way Tables

When two categorical variables are under consideration (or one variable from several different populations), a **two-way table** or **contingency table** is created. The table shows the distribution of data for each variable.

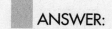 **EXAMPLE:** A survey of 12 adults asked, "Of chocolate, strawberry, and vanilla, which is your favorite ice cream flavor?" Each respondent's gender was also recorded. The results were as follows:

Flavor	Gender	Flavor	Gender
vanilla	male	chocolate	female
chocolate	female	chocolate	male
vanilla	male	vanilla	female
vanilla	male	strawberry	male
strawberry	female	chocolate	female
vanilla	male	vanilla	male

Create a two-way table that summarizes these data.

ANSWER:

		Flavor			
		Chocolate	Strawberry	Vanilla	Total
Gender	Female	3	1	1	5
	Male	1	1	5	7
	Total	4	2	6	12

Frequencies for each cell are called **joint frequencies** or the **joint distribution**. Two-way tables may or may not have a total row and a total column. It is recommended that they be included. These totals are called **marginal frequencies** or the **marginal distribution**.

Two-way (or contingency) tables can also display relative frequencies. These relative frequencies can be of the entire table, of rows, or of columns.

■ **EXAMPLE:** Create a two-way table of relative frequencies, with each cell being the relative frequency of the entire table. Round each relative frequency to three decimal places.

■ **ANSWER:**

Relative Frequency of the table		Flavor			
		Chocolate	Strawberry	Vanilla	Total
Gender	Female	$\frac{3}{12} = 0.250$	$\frac{1}{12} \approx 0.083$	$\frac{1}{12} \approx 0.083$	$\frac{5}{12} \approx 0.417$
	Male	$\frac{1}{12} \approx 0.083$	$\frac{1}{12} \approx 0.083$	$\frac{5}{12} \approx 0.417$	$\frac{7}{12} \approx 0.583$
	Total	$\frac{4}{12} \approx 0.333$	$\frac{2}{12} \approx 0.167$	$\frac{6}{12} = 0.500$	$\frac{12}{12} = 1$

OR

Relative Frequency of the table		Flavor			
		Chocolate	Strawberry	Vanilla	Total
Gender	Female	25.0%	8.3%	8.3%	41.7%
	Male	8.3%	8.3%	41.7%	58.3%
	Total	33.3%	16.7%	50.0%	100%

DIDYOUKNOW?

According to the International Ice Cream Association, the five most popular ice cream flavors are vanilla (29%), chocolate (8.9%), butter pecan (5.3%), strawberry (5.3%), and Neapolitan (4.2%).

 **EXAMPLE:** Create a two-way table of relative frequencies, with each cell being the relative frequency of the row. Round each relative frequency to three decimal places.

ANSWER:

Relative Frequency of the row		Flavor			
		Chocolate	Strawberry	Vanilla	Total
Gender	Female	$\frac{3}{5} = 0.600$	$\frac{1}{5} = 0.200$	$\frac{1}{5} = 0.200$	$\frac{5}{5} = 1$
	Male	$\frac{1}{7} \approx 0.143$	$\frac{1}{7} \approx 0.143$	$\frac{5}{7} \approx 0.714$	$\frac{7}{7} = 1$
	Total	$\frac{4}{12} \approx 0.333$	$\frac{2}{12} \approx 0.167$	$\frac{6}{12} = 0.500$	$\frac{12}{12} = 1.$

OR

Relative Frequency of the column		Flavor			
		Chocolate	Strawberry	Vanilla	Total
Gender	Female	75.0%	50.0%	16.7%	100%
	Male	25.0%	50.0%	83.3%	100%
	Total	100%	100%	100%	100%

The relative frequencies for each cell are called **conditional frequencies** or the **conditional distribution**. In this case, it is the conditional distribution for gender. For example, given that a respondent has the condition *gender is female*, the percentages that like chocolate, strawberry, or vanilla are 60%, 20%, and 20%, respectively.

■ **EXAMPLE:** Create a two-way table of relative frequencies, with each cell being the relative frequency of the column. Round each relative frequency to three decimal places.

■ **ANSWER:**

Relative Frequency of the column		Flavor			
		Chocolate	Strawberry	Vanilla	Total
Gender	Female	$\frac{3}{4} = 0.750$	$\frac{1}{2} = 0.500$	$\frac{1}{6} \approx 0.167$	$\frac{5}{12} \approx 0.417$
	Male	$\frac{1}{4} = 0.250$	$\frac{1}{2} = 0.500$	$\frac{5}{6} \approx 0.833$	$\frac{7}{12} \approx 0.583$
	Total	$\frac{4}{4} = 1$	$\frac{2}{2} = 1$	$\frac{6}{6} = 1$	$\frac{12}{12} = 1$

OR

Relative Frequency of the column		Flavor			
		Chocolate	Strawberry	Vanilla	Total
Gender	Female	75.0%	50.0%	16.7%	100%
	Male	25.0%	50.0%	83.3%	100%
	Total	100%	100%	100%	100%

This table shows the conditional distribution for flavor. For example, given that a respondent has the condition *preferred flavor is chocolate*, the percentages by gender are 75% female and 25% male.

The purpose of creating two-way tables with conditional distributions is to look for an **association** between two categorical variables of a single population, or one categorical variable of several populations. In the previous examples, one may be assessing whether there is an association between gender and preferred ice cream flavor.

Relative Frequency of the row		Flavor			
		Chocolate	Strawberry	Vanilla	Total
Gender	Female	60.0%	20.0%	20.0%	100%
	Male	14.3%	14.3%	71.4%	100%
	Total	33.3%	16.7%	50.0%	100%

In this instance, females are more likely to prefer chocolate to vanilla—60% to 20%. Males, on the other hand, prefer vanilla to chocolate—71.4% to 14.3%. The likelihood of preferring strawberry is about the same for either gender. There does seem to be an association between gender and flavor (or there are differences in flavor preference between males and females).

An association (or difference) such as this can be represented in a graphical form. The method of graphing a conditional distribution is with a **relative frequency segmented bar chart**. In a relative frequency segmented bar chart, one bar is drawn for each category of the conditional variable. Each bar is segmented into parts whose length is proportional to the percentage of categories of the other second variable.

EXAMPLE: Make a relative frequency segmented bar chart for the conditional distribution shown below.

Relative Frequency of the row		Flavor			
		Chocolate	Strawberry	Vanilla	Total
Gender	Female	60.0%	20.0%	20.0%	100%
	Male	14.3%	14.3%	71.4%	100%
	Total	33.3%	16.7%	50.0%	100%

ANSWER:

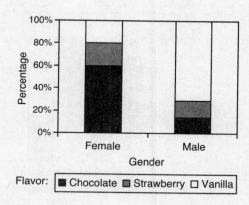

Note: The conditional variable is gender, so one bar exists for each category of the conditional variable. The segments correspond to the categories of the other variable: flavor.

It is clear from the graph that females prefer chocolate over vanilla and in a greater percentage than males. The same can be said of males preferring vanilla over chocolate and in a greater percentage than females. Strawberry was preferred about equally between the two genders.

EXAMPLE: Make a relative frequency segmented bar chart for the two-way table below, conditional on the variable *flavor*.

		Flavor			
		Chocolate	Strawberry	Vanilla	Total
Gender	Female	3	1	1	5
	Male	1	1	5	7
	Total	4	2	6	12

ANSWER: The conditional distribution on the variable flavor is shown below.

Relative Frequency of the column		Flavor			
		Chocolate	Strawberry	Vanilla	Total
Gender	Female	75.0%	50.0%	16.7%	100%
	Male	25.0%	50.0%	83.3%	100%
	Total	100%	100%	100%	100%

Its corresponding relative frequency segmented bar graph is

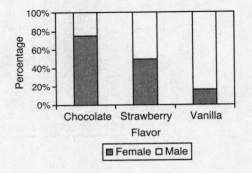

Note: The conditional variable is flavor, so one bar exists for each category of the conditional variable. The segments correspond to the categories of the other variable: gender.

It is clear from the graph that those who prefer chocolate tend to be female and those preferring vanilla tend to be male.

Time for a quiz
- Review strategies in Chapter 2
- Take Quizzes 1 and 2 at the REA Study Center
  (www.rea.com/studycenter)

Sampling and Experimentation

Overview of Methods of Data Collection

There are four methods of data collection with which you should be familiar: census, sample survey, experiment, and observational study.

A **census** is a study that observes, or attempts to observe, every individual in a population. The **population** is the collection of all individuals under consideration in the study.

A **sample survey** is a study that collects information from a sample of a population in order to determine one or more characteristics of the population. A **sample** is a selected subset of a population from which data are gathered.

An **experiment** is a study where the researcher deliberately influences individuals by imposing conditions and determining the individuals' responses to those conditions.

An **observational study** attempts to determine relationships between variables, but the researcher imposes no conditions like an experiment. Surveys are a form of observational study.

Sample surveys and experiments are discussed in greater detail later in this chapter.

EXAMPLE: The principal of a new school wants to know what mascot the incoming students would prefer. He summons each of the student body into his office and asks what the new mascot should be. What type of study is this?

ANSWER: This is a census. The principal has collected data from every individual in the population, in this case, the population of all school students.

DIDYOUKNOW?

The U.S. Constitution mandates that a national census, or population count, must be taken every ten years. Over time, the Census Bureau has begun collecting a great deal of additional information to learn about the characteristics of the U.S. population.

EXAMPLE: The principal of the new school is also interested in what type of dress code the school should have. He telephones the parents of every tenth student summoned to his office and asks their opinions about school dress codes. What type of study is this?

ANSWER: This is a sample survey. The principal has collected data from only a subset of the population, in this case, the population of all school parents.

EXAMPLE: A pharmaceutical company wishes to test a new medication it thinks will reduce cholesterol. A group of 20 volunteers is formed and each has his or her cholesterol level measured. Half is randomly assigned to take the new drug and the other half is given a dummy pill, a pill with no active ingredients. After 6 months the volunteers' cholesterol is measured again and any change from the beginning of the study recorded. What type of study is this?

ANSWER: This is an experiment. The company is imposing conditions (new drug, dummy pill) on the participants in the study to determine their response (change in cholesterol) to the conditions.

EXAMPLE: A health studies research lab is interested in the effect of certain vegetables on cholesterol level. A group of 20 volunteers is formed and each keeps a diary of his or her food consumption for the next 6 months, after which time the diaries are collected and cholesterol level is measured. The researchers then examine the relationship between the rate of consumption of certain vegetables and cholesterol level. What type of study is this?

ANSWER: This is an observational study. The researchers are imposing no conditions on the participants in the study.

TEST TIP

You'll have 90 minutes to complete 40 multiple-choice questions during the first section of the AP Statistics Exam. Plan to spend about 2 minutes and 15 seconds evaluating and responding to each multiple-choice item in order to have enough time to comfortably complete the entire section.

Planning and Conducting Surveys

The purpose of a sample survey is to determine one or more characteristics about a population from the results of the sample. The characteristic of the population is a **population parameter**. The result of the sample survey used to estimate the parameter is a **sample statistic**. There are many methods to obtain a sample from a population.

Sampling Methods

A **voluntary response sample** is composed of individuals who choose to respond to a survey because of interest in the subject, particularly those with strong opinions or attraction.

A **convenience sample** is composed of individuals who are easily accessed or contacted.

EXAMPLE: A journalism class prints a survey in their school newspaper. Readers are asked to clip the survey from the paper, complete it, and return it to a drop box in the school cafeteria. What type of sample is this?

ANSWER: This is a voluntary response sample. It is likely that only students interested in the survey's subject matter will take the time to respond.

EXAMPLE: A journalism class stations pollsters in front of the stadium during a football game. They ask each student who enters his or her opinion of the quality of the school's athletics program. What type of sample is this?

ANSWER: This is a convenience sample. The site was likely chosen because of the ease of gathering data there.

A **random sample (or probability sample)** is composed of individuals selected by chance. There are several types of random samples.

A **simple random sample (SRS)** is a sample where n individuals are selected from a population in a way that every possible combination of n individuals is equally likely. In general, one may think of an SRS as having all individuals of the population in a hat and selecting n individuals, without replacement, from the hat.

Note: Not only is it required of a simple random sample that each *individual* has an equally likely chance of being chosen, but also each possible group must be equally likely. That is, if an SRS is to have a size of $n = 5$, no possible group of five individuals can be any more or less likely to be selected than any other group.

A **stratified random sample** is a sample in which simple random samples are selected from each of several homogeneous subgroups of the population, known as **strata** (singular, **stratum**).

Note: In a stratified random sample, every individual in each subgroup has an equally likely chance of being chosen, but every individual in the population may not. Additionally, each possible group of individuals from the population will not have the same chance of being chosen.

A **cluster sample** is a sample in which a simple random sample of heterogeneous subgroups of the population is selected. The subgroups are known as **clusters**. The selected clusters may also themselves be subject to random sampling, or a census done for each.

A **multistage sample** is a sample resulting from multiple applications of cluster, stratified, and/or simple random sampling.

A **systematic random sample** is a sample where every *k*th individual is selected from a list or queue. The first selection is randomly chosen from the first *k* individuals.

EXAMPLE: A media research firm is conducting a poll on an upcoming election for city council. The firm obtains a list of all 15,000 registered voters in the council ward under consideration. The voters' ID numbers are entered into a computer and 500 are chosen at random, without replacement, to comprise the sample. What type of sampling design is this?

ANSWER: This is a simple random sample. Individuals are chosen at random, without replacement, from a list of the entire population. This is a "select n from a hat"-type of sampling.

Note: In this scenario, every possible group of 500 individuals has an equally likely chance of being selected, as does every individual.

EXAMPLE: A media research firm is conducting a poll on an upcoming election for city council. The firm obtains a list of all 15,000 registered voters in the council ward under consideration. The voters' ID numbers and party affiliations (Democrat, Republican, Independent) are entered into a computer. The firm randomly selects, without replacement, 200 Democrats, 200 Republicans, and 100 Independents to comprise the sample. What type of sampling design is this?

ANSWER: This is a stratified random sample. Individuals were first placed in strata, in this case, party affiliation. Then, simple random samples from each stratum were selected to make up the final sample.

Note: In this scenario, all individuals within their respective strata have an equally likely chance of being chosen, but all individuals in the population may not. If there are more Democrats than Republicans, then it is more likely that a particular Republican voter will be in the sample than a particular Democrat voter. Additionally, every possible group of 500 individuals from the population does not have an equally likely chance of being chosen, since samples of 500 Democrats, or 250 Democrats and 250 Republicans, for example, are not possible.

DIDYOUKNOW?

According to a 2009 poll, the District of Columbia, Rhode Island, and Hawaii have the highest proportions of Democrats. Utah, Wyoming, and Idaho have the highest proportions of Republicans.

EXAMPLE: A media research firm is conducting a poll on an upcoming election for city council. The firm obtains a list of all 15,000 registered voters in the council ward under consideration. The voters' ID numbers and voting precinct numbers (01–75) are entered into a computer. The firm first randomly selects 20 of the voting precincts. Then, from each of the chosen precincts, 25 voters are chosen with simple random samples. What type of sampling design is this?

ANSWER: This is a cluster sample. Individuals were first grouped into clusters by voting precinct, of which several were selected. Then, simple random samples from each cluster were selected to make up the final sample.

Note: An additional step could have been added to the previous example. If voters in the selected precincts were then grouped by party affiliation, and an SRS taken from each group, the result would have been a multistage sample.

EXAMPLE: A media research firm is conducting a poll on an upcoming election for city council. The firm obtains a list of all 15,000 registered voters in the council ward under consideration. The list is sorted by voter ID number. The firm randomly selects one person from the first 100 on the list; then it selects every 100th person after that. The sample size is 150. What type of sampling design is this?

ANSWER: This is a systematic random sample. One of the first k individuals on the list is randomly selected—in this case k 5 100—and every kth individual on the list is selected after that.

Note: Even though each individual has an equally likely chance of being selected in the systematic random sample, it is *not* equivalent to a simple random sample.

Each of the described designs has advantages and disadvantages.

The advantage of convenience samples and voluntary response samples is that gathering data is easy. Respondents are either close at hand, in the case of a convenience sample, or a portion of the initiative in data collection is up to the respondent, in the case of a voluntary response sample. Both of these can be poor designs, as they are subject to bias.

Bias and Sampling Error

Bias is the term for systematic deviation from the truth, in this case, the characteristic of the population (parameter) one wishes to estimate from the sample (statistic). A sampling method is biased if it tends to produce samples that do not represent the population. There are several terms used to describe bias.

Undercoverage occurs when some individuals of a population are not included in the sampling process.

Voluntary response bias is present in voluntary response samples. It occurs because people with strong opinions or interest in the survey topics tend to respond more frequently. Voluntary response bias is a form of undercoverage.

EXAMPLE: A journalism class prints a survey in their school newspaper about the school charging students for parking. Readers are asked to clip the survey from the paper, complete it, and return it to a drop box in the school cafeteria. Why is this sample biased?

ANSWER: This survey is subject to voluntary response bias, as it is likely that only students who drive to school now, or will drive soon, will take the time to respond. Nondrivers will tend to leave themselves out. Even if the population of interest were only driving students, those opposed to parking fees would be more likely to respond.

EXAMPLE: A journalism class stations pollsters in front of the stadium during a football game. They ask each student who enters his or her opinion of the quality of the school's athletics program. Why is this sample biased?

ANSWER: This convenience sample is subject to undercoverage. Only those students attending the football game will be able to register their opinions, potentially leaving out a large portion of the population.

Random sampling, when done properly, tends to reduce the effects of undercoverage by giving each individual in the population a chance to be selected for the survey. This is its greatest advantage—it helps produce samples that resemble the population. *Random sampling is a necessary part of a well-designed survey or sampling procedure.*

Simple random samples are, in practice, very difficult to produce. The advantage of cluster sampling and multistage sampling is that using these methods it can be easier to obtain a sample than using a simple random sample. In the previous example about the polling firm, were the surveys being conducted door to door, it would be easier to go to several houses in a few randomly selected voting precincts than to many houses scattered about the entire council ward. This saves resources—time and potentially money.

Another problem with random samples is variability. Each possible SRS from a population will generate a different sample statistic. The differences among these statistics are called **sampling variability** or **sampling error**. Sampling variability is a natural part of the random sampling process and should not be confused with bias. It is not the result of a poorly designed study.

The advantage of stratified random sampling is that it reduces sampling variability if there are differences between the strata in how they respond. It takes away the chance that one may get a disproportionate number of individuals from a stratum in the sample than exist in the population.

Even surveys conducted by using properly executed random sampling techniques are still subject to bias. **Nonresponse bias** is the situation where an individual selected to be in the sample is unwilling, or unable, to provide data. **Response bias** occurs when because of the manner in which an interview is conducted, because of the phrasing of questions, or because of the attitude of the respondent, inaccurate data are collected.

DIDYOU**KNOW?**

In the United States, national polls typically sample just 1,000 to 1,500 people out of a total population of over 185 million. The average margin of error with such a sample size is about 3 percent.

EXAMPLE: A media research firm is conducting a poll on an upcoming election for city council. The firm obtains a list of all 15,000 registered voters in the council ward under consideration and a simple random sample of 500 voters is chosen. Over a 24-hour period, telephone calls are placed to the voters, with follow-up calls made to those voters who do not answer the phone on the first attempt. Pollsters could not reach 37 of the 500 voters selected. What sources of bias could exist in this survey?

ANSWER: This survey could be subject to nonresponse bias. Thirty-seven voters could not be reached. Their opinions could affect the results of the survey, and the data gathered may not accurately represent the views of the population.

EXAMPLE: A journalism class conducts a simple random sample of students at their school. They ask each student, "Given the fact that our school has won seven championships in the last five years, do you favor or oppose reducing funding for athletic programs?" What sources of bias could exist in this survey?

ANSWER: The question is clearly leading the respondent to answer in opposition to reducing funding because of its mention of the school's recent athletic prowess. This is a form of response bias. A better question would be to ask, "Do you favor or oppose reducing funding for the school's athletic programs?"

Sampling with Random Digit Tables

A random digit table can be used to select a random sample from a population. Each member of the population is assigned an identification number, with the length of the number in digits determined by the size of the population. For populations of 10 or less, one digit is needed. Populations of 11–100 require two-digit numbers. Three digits are required for populations of size 101–1,000, and so on. Note that if one has 100 individuals in a population, assigning 01–100 is *not* correct. The number 100 has three digits and all of the others from 01 to 99 have two. However, 00 is a two-digit number that will work for 100.

Random numbers of the same length are selected from the random digit table by reading across the lines of the table. Each consecutive group of digits will either select one of the individuals in the population, or be ignored because it has already been selected, if sampling without replacement, or because it does not correspond with any member of the population.

EXAMPLE: Select a simple random sample of size $n = 6$ from the population of names below using the random digit table provided:

Anita	Billy	Carol	Doug	Elmer
Francine	Glenda	Hector	Ivy	Jose
Kelly	Lynn	Melvin	Nicole	Olive
Paul	Quincy	Rae	Sue	Tom

0 1 4 0 1	4 9 9 1 3	2 0 1 3 4	9 6 0 1 0
1 6 2 9 0	3 3 8 4 3	9 5 9 4 5	0 4 8 3 4
3 7 5 2 0	9 3 0 1 5	9 3 6 1 5	0 3 4 1 3

ANSWER: First, assign two-digit identification numbers to the members of the population.

01 Anita	02 Billy	03 Carol	04 Doug	05 Elmer
06 Francine	07 Glenda	08 Hector	09 Ivy	10 Jose
11 Kelly	12 Lynn	13 Melvin	14 Nicole	15 Olive
16 Paul	17 Quincy	18 Rae	19 Sue	20 Tom

Next, take consecutive two-digit numbers from the random number table, excluding values 21–99 and 00, and those that have already been selected.

0 1 4 0 1	4 9 9 1 3	2 0 1 3 4	9 6 0 1 0
1 6 2 9 0	3 3 8 4 3	9 5 9 4 5	0 4 8 3 4
3 7 5 2 0	9 3 0 1 5	9 3 6 1 5	0 3 4 1 3

The first five two-digit numbers (single underline) are 01, 40, 14, 99, and 13. These correspond to Anita, no selection, Nicole, no selection, and Melvin. So far, three of the sample have been selected. Continuing, the next several two-digit numbers (double underline) are as follows: 20—Tom; 13—Melvin, but he is a repeat and is not selected again; 49—no selection; 60—no selection; 10—Jose; and (skipping to the next line) 16—Paul.

The simple random sample of size $n = 6$ is Anita, Nicole, Melvin, Tom, Jose, and Paul.

Planning and Conducting Experiments

Recall that an **experiment** is a study where the researcher deliberately influences individuals by imposing conditions and determining the individuals' responses to those conditions. The individuals in an experiment are referred to as **experimental units**, or if they are people, **subjects**. Experimental units could be people, animals, plots of land, batteries, or any number of things on which conditions are imposed.

In an experiment, the researcher is interested in the response variable and how it is related to one or more explanatory variables called **factors**. Each factor has one or more **levels**, different quantities or categories of the factor. Combinations of different levels of the factors are called **treatments**. Sometimes, human subjects are given a faux treatment, known as a **placebo**, that resembles the real treatments under consideration.

EXAMPLE: A pharmaceutical company wishes to test a new medication it thinks will reduce cholesterol. A group of 20 volunteers is formed and each has his or her cholesterol level measured. Half is randomly assigned to take the new drug and the other half is given a placebo. After 6 months the volunteers' cholesterol is measured again and any change from the beginning of the study recorded. Identify the experimental units, factors and their levels, treatments, and response variable in this experiment.

ANSWER: The experimental units (subjects) in this study are the 20 volunteers. There is one factor, the medication, and it has two levels, the active pill and the placebo. Thus, there are only two treatments, the active pill and the dummy pill. The response variable is the change in cholesterol over the period of the study.

DIDYOUKNOW?

The U.S. Food and Drug Administration requires any new drug to undergo four phases of clinical trials. On average, the process of developing and testing a new drug takes over a decade and costs more than $500 million.

A diagram can assist visualizing the experiment's design.

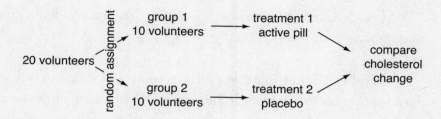

■ **EXAMPLE:** An agricultural researcher is interested in determining how much water and fertilizer are optimum for growing a certain plant. Twenty-four plots of land are available to grow the plant. The researcher will apply three different amounts of fertilizer (low, medium, and high) and two different amounts of water (light and heavy). These will be applied at random in equal combination to each of four plots. After 6 weeks, the plants' heights in each plot will be recorded. Identify the experimental units, factors and their levels, treatments, and response variable in this experiment.

ANSWER: The experimental units in this study are plots of land. There are two factors, fertilizer and water. Fertilizer has three levels: low, medium, and high. Water has two levels: light and heavy. There are total six treatments of fertilizer–water combinations: low–light, low–heavy, medium–light, medium–heavy, high–light, and high–heavy. The response variable is the height of the plants at the end of the study.

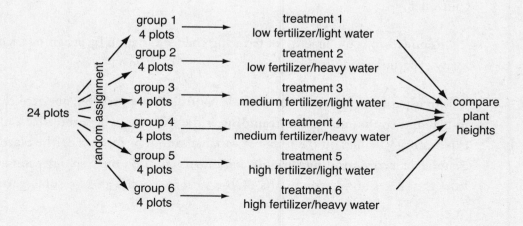

Characteristics of a Well-Designed Experiment

A well-designed comparative experiment has the following characteristics: control, randomization, and replication.

Control is the principle that potential sources of variation due to variables not under consideration must be reduced. These other variables are often called **lurking variables**. Control is achieved by making the experimental conditions as identical as possible for all experimental units.

One form of control in a comparative experiment is creating a baseline group or **control group**. A control group may be given no treatment, a faux treatment like a placebo, or an accepted treatment that is to be compared to another. In studies involving human subjects, controls are important to reduce a phenomenon known as the **placebo effect**, where subjects show a response to a treatment merely because the treatment is imposed regardless of its actual effect.

Another type of control is **blinding**. Blinding is the practice of denying knowledge to subjects of which treatment is being imposed upon them. This reduces the chances that the subjects will alter their behavior and introduce unwanted variability into the response. Sometimes, those collecting data in the experiment can introduce variability into the response, particularly when the response variable has a degree of subjectivity. When both subjects and evaluators are ignorant about which treatment a subject received, the study is called **double-blind**.

Randomization is the process by which treatments are assigned by a chance mechanism to the experimental units. It "averages out" variation due to variables that cannot be controlled.

Replication is the practice of reducing chance variation by assigning each treatment to many experimental units.

An experiment that does not include control, randomization, and replication is subject to bias and confounding. **Confounding** is the situation where the effects of two or more explanatory variables on the response variable cannot be separated. The placebo effect is an example of a **confounding variable**, because its effect on the response cannot be untangled from the effects of the treatment(s). Observational studies are also subject to confounding.

EXAMPLE: A pharmaceutical company wishes to test a new medication it thinks will reduce cholesterol. A group of 20 volunteers is formed and each has his or her cholesterol level measured. Half is randomly assigned to take the new drug and the other half is given a placebo. Neither group knows which pill it is taking. After 6 months the volunteers' cholesterol is measured again and any change from the beginning of the study recorded. Explain where control, randomization, and replication are present in this study.

ANSWER: Control is first implemented by having a control group—the group receiving the placebo. All subjects are alike in that all are taking a pill. The second form of control is by blinding the subjects. The subjects are less likely to introduce unwanted variability by changing their behavior, as they may if they knew which pill they were given. Randomization is provided by the random assignment of the two treatments to the subjects. Replication is present in that each treatment is imposed 10 times. (More would be better.)

EXAMPLE: A health studies research lab is interested in the effect of certain vegetables on cholesterol level. A group of 20 volunteers is formed and each keeps a diary of his food consumption for the next 6 months, after which time the diaries are collected and cholesterol level is measured. The researchers then examine the relationship between the rate of consumption of certain vegetables and cholesterol level. What confounding variables could be present in this observational study?

ANSWER: Sources of confounding could be exercise, diet outside of the vegetables being considered, or other health-related variables. Those who eat more vegetables may eat lower-fat diets, be less likely to smoke, or more likely to exercise, all variables that affect cholesterol level.

TEST TIP

During the AP Exam, you will have a break of just five minutes between the multiple-choice and free-response sections. Although you cannot eat or drink anything during the test, you may bring along a snack or a bottle of water for your break time.

Experiment Designs

There are two major types of experimental designs. The first is a **completely randomized design**. A completely randomized design is one in which all experimental units are assigned treatments solely by chance. No grouping of experimental units is done prior to assignment of treatments.

The second major design is the **randomized block design**. If a researcher has reason to believe that subgroups of the experimental units will respond differently to treatments because of some characteristic, the units are sorted into those subgroups before treatments are assigned. In an experiment, these subgroups are called **blocks**. Once units are assigned to blocks, treatments are randomly assigned to the units in each block. Blocking is a form of control to reduce unwanted variability in the response variable due to some variable other than the treatment(s). Experimental units may be sorted into blocks by one or more characteristics.

When only two treatments exist, experimental units are sometimes placed into pairs. These pairs may be units related by some variable, or may be single unit that receives each treatment at different times (known as reuse). This is a form of block design called **matched pairs**. If the pair consists of two experimental units, one is randomly assigned one of the two treatments with second unit receiving the other. If the pair is the same unit to be reused, one treatment is randomly assigned first, completed, and then the other treatment follows.

(Note: Do not confuse **blocking** and **stratification**. They perform similar functions in experiments and sampling, respectively, but blocks are part of an experimental design and strata are part of a sampling process.)

EXAMPLE: A pharmaceutical company wishes to test a new medication it thinks will reduce cholesterol. A group of 20 volunteers is formed and each has his or her cholesterol level measured. Half is randomly assigned to take the new drug and the other half is given a placebo. Neither group knows which pill it is taking. After 6 months the volunteers' cholesterol is measured again and any change from the beginning of the study recorded. What type of experimental design is this?

ANSWER: This is a completely randomized design.

EXAMPLE: A pharmaceutical company wishes to test a new medication it thinks will reduce cholesterol. A group of 20 volunteers is formed and each has his or her cholesterol level measured. After 6 months the volunteers' cholesterol is measured again and any change from the beginning of the study recorded. The researcher believes that regular exercise may influence the change in cholesterol level. Create a randomized block design that takes account of subjects who exercise regularly.

ANSWER: First, divide the subjects into two blocks: those who exercise regularly and those who do not. Then, randomly assign the treatments within each block, with half of each block receiving each treatment. A diagram is given as follows.

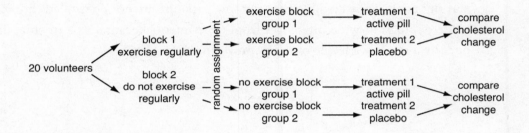

EXAMPLE: A pharmaceutical company wishes to test a new medication it thinks will reduce cholesterol. A group of 20 volunteers is formed and each has his or her cholesterol level measured. After 6 months the volunteers' cholesterol is measured again and any change from the beginning of the study recorded. The researcher believes that initial cholesterol level may influence the change in cholesterol level. Create a matched pairs design that takes account of subjects' initial cholesterol level.

ANSWER: Pair the subjects by initial cholesterol level, making the two with the highest levels a pair, then the third and fourth highest levels the next pair, and so on. Randomly assign one of each pair one of the treatments, with the other subject in the pair receiving the other treatment.

Generalizability of Results and Types of Conclusions that can be Drawn from Observational Studies, Experiments, and Surveys

When individuals are selected at random from a population, inferences to the population can be made. Without this random selection, inference is inappropriate. This typically applies to observational studies and surveys.

When treatments are randomly assigned to groups, cause–effect relationships between explanatory and response variables can be made. Without this random assignment, causal relationships cannot be drawn. This typically applies to experiments.

It is difficult in practice to draw cause–effect relationships from observational studies and surveys because possible confounding variables are not controlled. It is also difficult to make population-wide inferences from experiments because experimental units are infrequently a random sample of the population.

Time for a quiz
- Review strategies in Chapter 2
- Take Quizzes 3 and 4 at the REA Study Center
  (www.rea.com/studycenter)

Take Mini-Test 1
on Chapters 3–4
Go to the REA Study Center
(www.rea.com/studycenter)

Chapter 5

Anticipating Patterns

Probability

Probability describes the chance that a certain outcome of a random phenomenon will occur. **Random phenomena** are those where their outcomes are unpredictable in the short term, but a long-term pattern emerges. One can determine the probability of a particular event of a random phenomenon by considering a set of equally likely outcomes or by examining its long-term relative frequency. In either case, the probability of an event is always a value between 0 and 1. This is noted as $0 \leq P(\text{Event}) \leq 1$.

The probability of an event given *equally likely* outcomes is determined by the formula

$$\text{probability of event } A = P(A) = \frac{\text{number of successful outcomes}}{\text{total number of possible } equally \ like}$$

EXAMPLE: A small high school has 16 freshmen, 15 sophomores, 12 juniors, and 17 seniors. If a student from the school is selected at random, what is the probability that the student is a junior?

ANSWER: When choosing a student at random, there are $16 + 15 + 12 + 17 = 60$ total equally likely outcomes of which 12 are juniors (a successful outcome). Therefore, $P(\text{junior}) = \dfrac{12}{60} = 0.2 \cdot$

Probabilities of events can also be determined by examining the long-term relative frequency of their occurrence. The principle behind this is the **law of large numbers**. The law of large numbers states: The long-term relative frequency of an event gets closer to the true relative frequency as the number of trials of the random phenomenon increases.

EXAMPLE: A thumbtack is tossed onto the floor. What is the probability of it landing with its point up?

ANSWER: Even though there are two outcomes, "point up" and "point down," they may not be equally likely. However, one could approximate the true probability by tossing a tack many times and determining its long-term relative frequency. The graph below indicates that the probability of a tack landing "point up" is about 0.42.

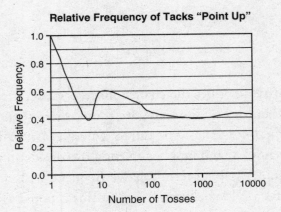

The **complement** of an event is an event *not* occurring. The following formula is used to describe complementary events.

For **complementary events:** $P(\text{not } A) = 1 - P(A)$

EXAMPLE: A thumbtack is tossed onto the floor. What is the probability of it *not* landing with its point up?

ANSWER: From the previous example we determined that $P(\text{point up}) = 0.42$. Thus, $P(\text{not point up}) = 1 - P(\text{point up}) = 1 - 0.42 = 0.58$.

On occasion, two (or more) events from a random phenomenon are considered at the same time. If those events cannot occur simultaneously, we call them **disjoint** (or **mutually exclusive**). For example, when considering a randomly selected high school student, the events "freshman" and "sophomore" are disjoint since the student cannot be both a freshman and a sophomore. The events "boy" and "sophomore" are not disjoint because the student could be both a sophomore and a boy.

The **addition rule** for probability aids in computing the chances of one or several events occurring at a given time.

Addition rule: $P(A \cup B) = P(A) + P(B) - P(A \cap B)$. (Note: $\cup$ is the symbol for *union* or the word "or"; $\cap$ is the symbol for *intersection* or the word "and.") If events A and B are disjoint, $P(A \cap B) = 0$.

EXAMPLE: The table below shows the breakdown of students in a small school by class and gender. What is the probability that a randomly selected student will be a sophomore or a junior?

	Freshmen	Sophomores	Juniors	Seniors	Total
Boys	9	7	5	7	28
Girls	7	8	7	10	32
Total	16	15	12	17	60

ANSWER: $P(\text{sophomore}) = \dfrac{15}{60}$, $P(\text{junior}) = \dfrac{12}{60}$, so,

$$P(\text{sophomore} \cup \text{junior}) = P(\text{sophomore}) + P(\text{junior}) = \frac{15}{6} + \frac{12}{60} = \frac{27}{60} = 0.45.$$

EXAMPLE: What is the probability that a randomly selected student will be a sophomore or a girl?

ANSWER: $P(\text{sophomore}) = \dfrac{15}{60}$, $P(\text{girl}) = \dfrac{32}{60}$, so,

$$P(\text{sophomore} \cup \text{girl}) = P(\text{sophomore}) + P(\text{girl}) - P(\text{sophomore} \cap \text{girl})$$
$$= \frac{15}{60} + \frac{32}{60} - \frac{8}{60} = \frac{39}{60} = 0.65.$$

A **Venn diagram** is a graphical representation of sets or outcomes and how they intersect. A Venn diagram for two events A and B is shown below. The black oval corresponds to the outcomes in event A; the light gray oval does the same for event B. The intersection is seen as dark gray. The outcomes belonging to neither event are in the white area.

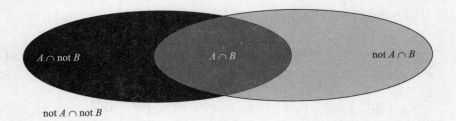

Consider the previous example. Sophomores are 15 of the students and are within the black oval—8 girls and 7 boys. Girls are 32 of the students and are in the light gray oval—8 sophomores and 24 from other classes. There are 8 sophomore girls, so there are 8 in the intersection. Twenty-one students are neither sophomores nor girls (freshman boys, junior boys, and senior boys). The $P(\text{sophomore} \cup \text{girl})$ is equal to the number of students that fall into either the sophomore or girl ovals or both, which is $7 + 8 + 24 = 39$, divided by the total number of students, which is 60.

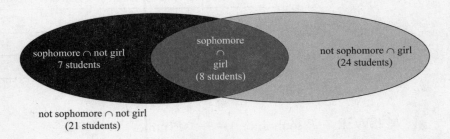

Sometimes we are interested in the probability of an event occurring given that another event has occurred. This is called **conditional probability**.

Conditional probability: $P(B \mid A) = \dfrac{P(A \cap B)}{P(A)}$. (Note: the vertical bar $\mid$ is the symbol for "given that" and precedes the condition that has occurred.)

■ EXAMPLE: What is the probability that a randomly selected student is a girl *given that* the student is a sophomore?

■ ANSWER: $P(\text{sophomore}) = \dfrac{15}{60}$, $P(\text{girl} \cap \text{sophomore}) = \dfrac{8}{60}$, so,

$$P(\text{girl} \mid \text{sophomore}) = \frac{P(\text{girl} \cap \text{sophomore})}{P(\text{sophomore})} = \frac{8/60}{} = \frac{8}{} \approx 0.533.$$

Another way to see this is that once it is established that the student is a sophomore, there are only 15 students it could possibly be and 8 of those are girls.

When two events are related, and the fact that one event has occurred changes the probability that the second event occurs, the events are called **dependent**. Therefore, if knowing that one event has occurred does *not* change the chance that the second event occurs, the events are called **independent**.

Independence check: Events A and B are independent if $P(B|A) = P(B)$.

■ EXAMPLE: Are the events "girl" and "sophomore" independent?

■ ANSWER: $P(\text{girl}) = \dfrac{32}{60} \approx 0.533$ and $P(\text{girl} \mid \text{sophomore}) = \dfrac{8/60}{15/60} = \dfrac{8}{15} \approx 0.533.$

Since $P(\text{girl}) = P(\text{girl}|\text{sophomore})$, the events "girl" and "sophomore" are independent.

■ EXAMPLE: Are the events "girl" and "junior" independent?

■ ANSWER: $P(\text{girl}) = \dfrac{32}{60} \approx 0.533$ and $P(\text{girl} \mid \text{junior}) = \dfrac{7/60}{12/60} = \dfrac{7}{12} \approx 0.583.$ Since $P(\text{girl}) \neq P(\text{girl}|\text{junior})$, the events "girl" and "junior" are *not* independent.

When we are interested in the probability of two events occurring simultaneously, or in successive trials, we use the **multiplication rule**.

Multiplication rule: $P(A \cap B) = P(A) \cdot P(B|A)$

EXAMPLE: Two students are selected at random from the school without replacement. What is the probability that they are both boys?

	Freshmen	Sophomores	Juniors	Seniors	Total
Boys	9	7	5	7	28
Girls	7	8	7	10	32
Total	16	15	12	17	60

ANSWER: $P(\text{first is boy} \cap \text{second is boy}) = P(\text{first is boy}) \cdot P(\text{second is boy}|\text{first is boy})$.

$P(\text{first is boy}) = \dfrac{28}{60}$ and $P(\text{second is boy} \mid \text{first is boy}) = \dfrac{27}{59}$, because the first boy is not replaced leaving 27 boys out of 59 students.

$P(\text{first is boy} \cap \text{second is boy}) = \dfrac{28}{60} \cdot \dfrac{27}{59} = \dfrac{756}{3,540} \approx 0.214$.

If events are **independent**, then $P(B|A) = P(B)$, so the **multiplication rule** becomes $P(A \cap B) = P(A) \cdot P(B)$.

EXAMPLE: Two fair six-sided dice are rolled. What is the probability that the first die shows an odd number and the second die shows a number greater than 4?

ANSWER: Since the first die's outcome does not influence the second's, the two events are independent. Thus,

$P(\text{first die odd} \cap \text{second die} > 4) = P(\text{first die odd}) \cdot P(\text{second die} > 4)$

$= \dfrac{3}{6} \cdot \dfrac{2}{6} = \dfrac{6}{36} \approx 0.167.$

A tree diagram is a helpful way to picture the probabilities of successive or compound events, particularly when there are conditional probabilities to consider. A tree diagram for two events, *A* and *B*, is shown below. Probabilities for each outcome are listed on the branches. The probability of the final result is the product of the probabilities along the branches.

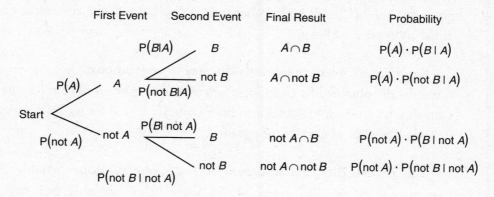

First Event	Second Event	Final Result	Probability
	$P(B|A)$ — B	$A \cap B$	$P(A) \cdot P(B | A)$
$P(A)$ — A	not B / $P(\text{not } B|A)$	$A \cap$ not B	$P(A) \cdot P(\text{not } B | A)$
Start	$P(B|\text{ not } A)$ — B	not $A \cap B$	$P(\text{not } A) \cdot P(B | \text{ not } A)$
$P(\text{not } A)$ — not A	not B / $P(\text{not } B | \text{ not } A)$	not $A \cap$ not B	$P(\text{not } A) \cdot P(\text{not } B | \text{ not } A)$

EXAMPLE: At a particular school 47% of the students are boys and 53% are girls. Of the boys, 25% are seniors. Of the girls, 31% are seniors. If a randomly selected student is a senior, what is the probability that it is a girl?

ANSWER: Making a tree diagram, let event *G* represent a girl and event *S* represent a senior.

First Event	Second Event	Final Result	Probability
	0.31 — S	$G \cap S$	$(0.53)(0.31) = 0.1643$
0.53 — G	not S / 0.69	$G \cap$ not S	$(0.53)(0.69) = 0.3657$
Start	0.25 — S	not $G \cap S$	$(0.47)(0.25) = 0.1175$
0.47 — not G	0.75 — not S	not $G \cap$ not S	$(0.47)(0.75) = 0.3525$

The question asks, what is $P(G|S)$?

$$P(G|S) = \frac{P(G \cap S)}{P(S)} = \frac{0.1643}{0.1643 + 0.1175} = \frac{0.1643}{0.2818} \approx 0.5830.$$

Discrete Random Variables

A **random variable** is a *numerical* outcome of a random phenomenon. It could be the number of heads seen in 10 flips of a coin, the number of pets in an animal shelter on any given day, the heights of high school volleyball players, or the amount of gasoline purchased at the most recent fill-up.

Random variables can be either **discrete** or **continuous**. Discrete random variables are those usually obtained by counting, such as counting heads of coin flips or pets in a shelter. Continuous random variables are those typically found by measuring, such as heights of volleyball players or amount of gasoline.

The **probability distribution** of a discrete random variable X is a function matching all n possible outcomes of the random variable (x_i) and their associated probabilities $P(x_i)$. For a probability distribution to be valid, it must meet two characteristics:

(1) For each value of x_i, where $i = 1, 2, \ldots, n$, $0 \leq P(x_i) \leq 1$.

(2) $\sum_{i=1}^{n} P(x_i) = 1.$

EXAMPLE: The number of cars X owned by the families of students at a small high school can be given by the probability distribution:

Number of cars, x	0	1	2	3	4
Probability, P(x)	0.05	0.15	0.35	0.30	?

What is the probability that a randomly selected student's family owns four cars?

ANSWER: Since the total of the probabilities must equal 1, the probability that student's family owns four cars is $1 - (0.05 + 0.15 + 0.35 + 0.30) = 1 - 0.85 = 0.15$.

EXAMPLE: The number of cars X owned by the families of students at a small high school can be given by the probability distribution:

Number of cars, x	0	1	2	3	4
Probability, P(x)	0.05	0.15	0.35	0.30	0.15

Create a probability histogram for the random variable X.

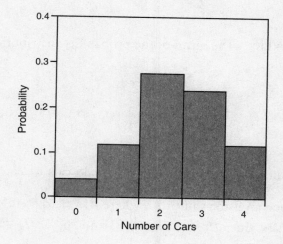

ANSWER: Measures of center and spread can be computed for discrete random variables just like data. The most common of these measures are mean, median, standard deviation, and variance.

Mean of a discrete random variable X: $\mu_X = \sum_{i=1}^{n} x_i \cdot P(x_i)$. The mean μ_X of random variable X is also referred to as the **expected value of X**, and symbolized $E(X)$.

Median of a discrete random variable X: The value of X such that $P(X \le x) \ge 0.5$ and $P(X \ge x) \ge 0.5$.

Variance of a discrete random variable X: $\sigma_X^2 = \sum_{i=1}^{n} (x_i - \mu_x)^2 \cdot P(x_i)$.

Standard deviation of a discrete random variable X: $\sigma_X = \sqrt{\sigma_X^2}$.

EXAMPLE: The number of cars X owned by the families of students at a large high school can be given by the probability distribution:

Number of cars, x	0	1	2	3	4
Probability, P(x)	0.05	0.15	0.35	0.30	0.15

Find the mean, median, variance, and standard deviation of the probability distribution.

ANSWER: The mean of the probability distribution is

$$\mu_X = \sum_{i=1}^{n} x_i \cdot P(x_i)$$
$$= (0)(0.05) + (1)(0.15) + (2)(0.35) + (3)(0.3) + (4)(0.15)$$
$$= 2.35 \text{ cars.}$$

This can also be written as $E(X) = 2.35$ cars.

The median is the value of X where $P(X \le x) \ge 0.5$ and $P(X \ge x) \ge 0.5$. This occurs at $X = 2$ cars, since $P(X \le 2) = 0.55$ and $P(X \ge 2) = 0.8$.

The variance of the probability distribution is

$$\sigma_X^2 = \sum_{i=1}^{n} (x_i - \mu_x)^2 \cdot P(x_i)$$
$$= (0 - 2.35)^2(0.05) + (1 - 2.35)^2(0.15) + (2 - 2.35)^2(0.35) + (3 - 2.35)^2$$
$$(3 - 2.35)^2(0.3) + (4 - 2.35)^2(0.15)$$
$$= 1.1275 \text{ cars}^2.$$

The standard deviation of the probability distribution is

$$\sigma_X = \sqrt{\sigma_X^2} = \sqrt{1.1275} \approx 1.0618 \text{ cars.}$$

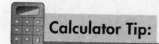
Calculator Tip:

The mean and standard deviation of a probability distribution can be calculated with the graphing calculator. Consider the previous example.

From the home screen, go to the list editor by pressing $\boxed{\text{STAT}}$; then from the **EDIT** menu choose **1:Edit...**. Enter the values of the random variable x_i in list L1 and enter the corresponding probabilities $p(x_i)$ in list **L2**.

L1	L2	L3 2
0	.05	------
1	.15	
2	.35	
3	.3	
4	.15	
------	------	

L2(5) =.15

Return to the home screen and press $\boxed{\text{STAT}}$, scroll right to the **CALC** menu, and then choose **1: 1–Var Stats**. Press $\boxed{\text{L1}}$ $\boxed{,}$ $\boxed{\text{L2}}$, and then press $\boxed{\text{ENTER}}$.

1—Var Stats L1, L
2█

The mean of the probability distribution will be listed as $\overline{\text{X}}$ (even though the correct symbol for it would be μ), and the standard deviation will be listed as σx. Note that the value of n should be 1, as the sum of the frequencies (probabilities) is 1.

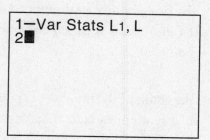

1—Var Stats
$\overline{x}$=2.35
Σx=2.35
Σx^2=6.65
Sx=
σx=1.061838029
↓n=1

Binomial Random Variables

A random variable X is called a **binomial random variable** if it meets the following conditions:

(1) There are a fixed number of trials, n, of a random phenomenon.

(2) There are only two possible outcomes on each trial, often called *success* and *failure*.

(3) The probability of success, p, is constant for each trial.

(4) Each trial is independent of the other trials.

The random variable X is the count of the number of successes in the n trials.

The probability that exactly k successes are obtained in n trials of a binomial random variable is $P(X=k)=\binom{n}{k}p^{k}(1-p)^{n-k}$, where $\binom{n}{k}$ is called the *binomial coefficient* and

has a value $\binom{n}{k}=\dfrac{n!}{k!(n-k)!}$.

EXAMPLE: The probability of a thumbtack landing "point up" when tossed is 0.42. If a thumbtack is tossed eight times, what is the probability that it lands "point up" exactly twice?

ANSWER: The count of the number of "point up" landings of a tossed tack is a binomial random variable because the four conditions listed above are satisfied. The probability of exactly two tosses of eight "point up" landings is

$$P(X=2)=\binom{8}{2}(0.42)^{2}(1-0.42)^{8-2}$$

$$=\frac{8!}{2!6!}(0.42)^{2}(0.58)^{6}$$

$$=28(0.42)^{2}(0.58)^{6}$$

$$\approx 0.188.$$

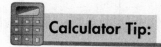

Calculator Tip:

Binomial probabilities can be calculated with the graphing calculator. Consider the previous example.

From the home screen, press ⟦DISTR⟧ (which is ⟦2nd⟧ ⟦VARS⟧), and then choose A: binompdf(. This is the binomial probability density function command and is used for computing binomial probabilities for a particular number of successes.

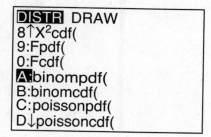

The syntax for the binompdf command is binompdf (number of trials, probability of success, number of successes). Key in 8, .42, 2) and then press ⟦ENTER⟧.

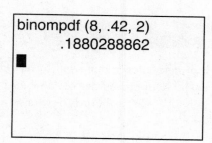

The binomial probability for getting exactly two successes in eight trials with the probability of success of 0.42 is about 0.188.

EXAMPLE: The probability of a thumbtack landing "point up" when tossed is 0.42. If a thumbtack is tossed eight times, what is the probability that it lands "point up" at most twice?

ANSWER: The probability of at most two of eight "point up" landings is $P(X \leq 2) = P(X = 0) + P(X = 1) + P(X = 2)$.

From the previous example $P(X = 2) \approx 0.188$.

$$P(X = 1) = \binom{8}{1}(0.42)^1(1 - 0.42)^{8-1}$$
$$= 8(0.42)^1(0.58)^7$$
$$\approx 0.074.$$

$$P(X = 0) = \binom{8}{0}(0.42)^0(1 - 0.42)^{8-0}$$
$$= 1(0.42)^0(0.58)^8$$
$$\approx 0.013.$$

So, $P(X \leq 2) \approx 0.188 + 0.074 + 0.013 \approx 0.275$.

TEST TIP

Not sure if your calculator is approved for use on the AP Statistics Exam? You can check the most recent list of approved devices on the AP website at http://www.collegeboard.org. Calculators with QWERTY keyboards or wireless capabilities are never permitted.

Calculator Tip:

Cumulative binomial probabilities can be calculated with the graphing calculator. Consider the previous example.

From the home screen, press $\boxed{\text{DISTR}}$, and then choose **B: binomcdf(**. This is the binomial cumulative density function command and is used for computing binomial probabilities of a given number of successes or less.

```
DISTR DRAW
9↑Fpdf(
0:Fcdf(
A:binompdf(
B:binomcdf(
C:poissonpdf(
D:poissoncdf(
E↓geometpdf(
```

The syntax for the binomcdf command is binomcdf (number of trials, probability of success, number of successes or less). Key in **8, .42, 2)** and then press ENTER.

```
binomcdf (8, .42, 2)
          .2750234624
```

The binomial probability for getting two or fewer successes in eight trials with the probability of success of 0.42 is about 0.275.

The **binomial distribution** is the probability distribution of a binomial random variable, that is, all possible outcomes of X in n trials and their associated probabilities.

EXAMPLE: Give the binomial distribution of a thumbtack landing "point up" when it is tossed eight times and the probability of success is 0.42.

ANSWER: Using the calculator's binompdf function, we get

Number of "point up"	0	1	2	3	4	5	6	7	8
Probability	0.013	0.074	0.188	0.272	0.246	0.143	0.052	0.011	0.0001

EXAMPLE: Create a probability histogram for the binomial distribution of a thumb-tack landing "point up" when it is tossed four times and the probability of success is 0.42.

ANSWER: The probability distribution is

Number of "point up"	0	1	2	3	4
Probability	0.113	0.328	0.356	0.172	0.031

The corresponding histogram is

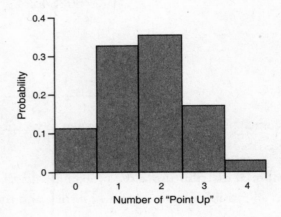

The mean and variance of a binomial random variable can be easily calculated. When the formulas $\mu_X = \sum_{i=1}^{n} x_i \cdot P(x_i)$ and $\sigma_X^2 = \sum_{i=1}^{n} (x_i - \mu_X)^2 \cdot P(x_i)$ are applied to a binomial random variable, the following result:

Mean of a binomial random variable X: $\mu_X = E(X) = np$.

Variance of a binomial random variable X: $\sigma_X^2 = np(1-p)$.

Standard deviation of a binomial random variable X: $\sigma_X = \sqrt{np(1-p)}$.

EXAMPLE: A thumbtack that lands "point up" with probability 0.42 is tossed eight times. What are the mean, variance, and standard deviation of the number of times it lands "point up"?

ANSWER:

The mean is $\mu_X = np = (8)(0.42) = 3.36$ times,

$$\sigma_X^2 = (8)(0.42)(0.58) = 1.9488 \text{ times}^2,$$

$\sigma_X = \sqrt{(8)(0.42)(0.58)} \approx 1.396$, times. (Note: The mean number of times the tack lands "point up" in eight tosses is 3.36. Even though the tack itself cannot land "point up" 3.36 times, it can on average.)

Geometric Random Variables

A random variable X is called a **geometric random variable** if it meets the following conditions:

(1) There are only two possible outcomes on each trial, often called *success* and *failure*.

(2) The probability of success, p, is constant for each trial.

(3) Each trial is independent of the other trials.

The random variable X is the count of the number of trials until the first success is obtained. A geometric random variable results from what is often called a "wait time" situation.

The probability that exactly k trials occur before the first success of a geometric random variable is obtained is $P(X = k) = p(1 - p)^{k-1}$.

EXAMPLE: The probability of a thumbtack landing "point up" when tossed is 0.42. What is the probability that the tack is tossed exactly four times before it lands "point up"?

ANSWER: The number of trials before a tossed tack lands "point up" is a geometric random variable because the three conditions listed above are satisfied. The probability of exactly four tosses before landing "point up" is

$$P(X = 4) = (0.42)(1 - 0.42)^{4-1}$$
$$= (0.42)(0.58)^3$$
$$\approx 0.082.$$

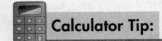 **Calculator Tip:**

Geometric probabilities can be calculated with the graphing calculator. Consider the previous example.

From the home screen, press $\boxed{\text{DISTR}}$, and then choose **E: geometpdf(**. This is the geometric probability density function command and is used for computing geometric probabilities for a particular number of trials.

```
DISTR DRAW
0↑Fcdf(
A:binompdf(
B:binomcdf(
C:poissonpdf(
D:poissoncdf(
E:geometpdf(
F:geometcdf(
```

The syntax for the geometpdf command is geometpdf(probability of success, number of trials). Key in **.42,4)** and then press $\boxed{\text{ENTER}}$.

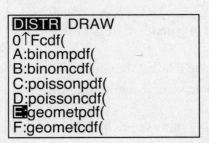

```
geometpdf(.42, 4)
              .08194704
```

The geometric probability for getting the first success on the fourth trial with the probability of success of 0.42 is about 0.082.

EXAMPLE: The probability of a thumbtack landing "point up" when tossed is 0.42. What is the probability that the tack is tossed at most four times before it lands "point up"?

ANSWER: The probability of at most four trials before landing "point up" is
$$P(X \le 4) = P(X = 1) + P(X = 2) + P(X = 3) + P(X = 4).$$

From the previous example, $P(X = 4) \approx 0.082$.

$$
\begin{aligned}
P(X = 1) &= (0.42)(1 - 0.42)^{1-1} \\
&= (0.42)(0.58)^0 \\
&\approx 0.42.
\end{aligned}
$$

$$
\begin{aligned}
P(X = 2) &= (0.42)(1 - 0.42)^{2-1} \\
&= (0.42)(0.58)^1 \\
&\approx 0.244.
\end{aligned}
$$

$$
\begin{aligned}
P(X = 3) &= (0.42)(1 - 0.42)^{3-1} \\
&= (0.42)(0.58)^2 \\
&\approx 0.141.
\end{aligned}
$$

So, $P(X \le 4) \approx 0.42 + 0.244 + 0.141 + 0.082 \approx 0.887$.

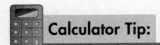

Calculator Tip:

Cumulative geometric probabilities can be calculated with the graphing calculator. Consider the previous example.

From the home screen, press $\boxed{\text{DISTR}}$, and then choose **F: geometcdf(**. This is the geometric cumulative density function command and is used for computing geometric probabilities of a given number of trials or less.

```
DISTR DRAW
0↑Fcdf(
A:binompdf(
B:binomcdf(
C:poissonpdf(
D:poissoncdf(
E:geometpdf(
F:geometcdf(
```

The syntax for the geometcdf command is geometcdf (probability of success, number of trials). Key in .42,4) and then press $\boxed{\text{ENTER}}$.

```
geometcdf(.42, 4)
                 .88683504

```

The geometric probability for getting the first success in four or fewer trials with the probability of success of 0.42 is about 0.887.

The **geometric distribution** is the probability distribution of a geometric random variable, that is, all possible outcomes of X before the first success is seen and their associated probabilities.

EXAMPLE: Give the geometric distribution of a thumbtack landing "point up" if the probability of success is 0.42. Create the corresponding probability histogram.

ANSWER: Using the calculator's geometpdf function, we get

Number of trials	1	2	3	4	5	6	7	8	...
Probability	0.42	0.244	0.141	0.082	0.048	0.028	0.016	0.009	...

The corresponding histogram is

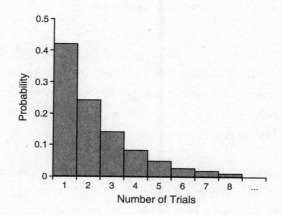

Note: that the geometric distribution goes on to infinity, but there will come a point where the number of trials before seeing the first success is practically zero.

The mean and variance of a geometric random variable can be easily calculated. When the formulas $\mu_X = \sum_{i=1}^{n} x_i \cdot P(x_i)$ and $\sigma_X^2 = \sum_{i=1}^{n} (x_i - \mu_X)^2 \cdot P(x_i)$ are applied to a geometric random variable, the following result (in spite of the number of possibilities being infinite):

Mean of a geometric random variable X: $\mu_X = E(X) = \dfrac{1}{p}$.

Variance of a geometric random variable X: $\sigma_X^2 = \dfrac{1-p}{p^2}$.

Standard deviation of a geometric random variable X: $\sigma_X = \sqrt{\dfrac{1-p}{p^2}}$.

EXAMPLE: A thumbtack which lands "point up" with probability 0.42 is tossed. What are the mean, variance, and standard deviation of the number of trials before it lands "point up"?

ANSWER: The mean is $\mu_X = \dfrac{1}{0.42} \approx 2.381$ times, $\sigma_X^2 = \dfrac{1-0.42}{0.42^2} \approx 3.288$ times^2,

$$\sigma_X = \sqrt{\dfrac{1-0.42}{0.42^2}} \approx 1.813 \text{ times.}$$

Simulations of Random Phenomena

Sometimes one is interested in finding the probability distribution of a random phenomenon, but the situation is too complex to use the methods outlined in this chapter thus far. Perhaps the scenario does not generate a binomial or geometric distribution. Computing the probabilities algebraically or with diagrams may be very time consuming. In these cases, **simulation** is a tool to answer the question.

Simulation is a method to model chance behavior that accurately mimics the situation being considered. The mechanism used to determine outcomes is called a **chance device**. A chance device could be a coin, a handful of dice, a random digit table, or computer/calculator code that generates random values. In this book, a **random digit table** will be used, as that is the chance device that will likely appear on the examination.

To conduct a simulation, follow these five steps:

(1) Describe the situation and its possible outcomes.

(2) Identify how a chance device used will model the outcomes of the situation.

(3) Describe what one trial of the experiment will be and what response variable will be recorded.

(4) Conduct several trials of the experiment.

(5) Analyze the distribution of the response variable and draw a conclusion.

EXAMPLE: Boxes of a cereal contain collector cards of characters from a movie. Some cards are more common than others. The distribution of the character cards in the boxes is given by the following:

Alfred the Agile: 20%

Brian the Brave: 30%

Craig the Crafty: 50%

What is the mean number of cereal boxes one would have to buy in order to get all three cards? Use the random digit table below to conduct a simulation. Do five trials of your simulation.

0 0 2 3 3	5 4 8 3 0	3 9 1 0 8
0 8 9 9 6	1 4 2 2 3	3 9 2 8 0
7 1 4 0 5	4 9 9 5 3	3 0 3 2 4

ANSWER: Conduct the simulation following the five-step process:

(1) Simulate the opening of a box of cereal and determine which of the three cards (A, B, C) is in the box.

(2) Random digits will be taken one at a time from the table, each representing one box of cereal. The value of the digit corresponds to the card in the box as follows: 0–1 = A, 2–4 = B, 5–9 = C.

(3) One trial will consist of selecting random digits until at least one result of each of "A," "B," and "C" is obtained. The response variable is the number of boxes of cereal opened.

(4)

	0	0	2	3	3	5	4	8	3	0	3	9	1	0	8
Card	A	A	B	B	B	C	B	C	B	A	B	C	A	A	C
No. of boxes						6 boxes				4 boxes			3 boxes		

	0	8	9	9	6	1	4	2	2	3	3	9	2	8	0
Card	A	C	C	C	C	A	B	B	B	B	B	C	B	C	A
No. of boxes							10 boxes							7 boxes	

7	1	4	0	5	4	9	9	5	3	3	0	3	2	4

(Note: When the end of the first line was reached, the trial was continued on the second line.)

(5) The numbers of boxes needed to get all three cards were 3, 4, 6, 7, and 10. The mean number of boxes is $\bar{x} = 6$.

DIDYOUKNOW?

Kellogg's put the first prizes in boxes of Corn Flakes in 1906. The prize, called the Funny Jungleland Moving Pictures Book, allowed users to create various combinations of the head, body, and legs of an elephant, alligator, giraffe, and lion using sliding pictures.

EXAMPLE: Fifteen students from a large high school have been selected to attend a meeting across town. Parents from the school have been asked to provide transportation to the event in their private cars and are to phone the school if interested. The school will accept calls until enough vehicles are obtained to transport the 15 students. The probability distribution below shows the number of passengers a car can hold, not including the driver, and the percentage of parents who own cars of that capacity.

Number of passengers	1	3	4	6
Percentage of cars	0.05	0.50	0.35	0.10

What is the average number of parents that will need to call to have enough transportation for 20 students? Use the random digit table below to conduct a simulation. Do five trials of your simulation, starting each trial on a new line.

```
01405    49953    30324
96080    51290    33843
95945    04834    37520
93015    93615    03413
36370    10356    03768
```

ANSWER: Conduct the simulation following the five-step process:

(1) Simulate the phone call of a parent and determine how many passengers the parent's car will hold.

(2) Random digits will be taken two at a time from the table, each representing one parent's car. The value of the digits corresponds to the number of passengers the car will hold: 01–05 = 1 passenger, 06–55 = 3 passengers, 56–90 = 4 passengers, 91–99 and 00 = 6 passengers.

(3) One trial will consist of selecting pairs of random digits and recording the number of passengers the car holds until a total of 15 or more passengers are obtained. The response variable is the number of parents that called.

(4)

```
                0  1│4  0│5         4│9  9│5  3│    3  0  3  2  4
No. of passengers  1     3        3         6    3
Total passengers   1     4        8        14   17
No. of calls                5
```

```
                9  6│0  8│0         5│1  2│9  0│    3  3  8  4  3
No. of passengers  6     3        1         3    4
Total passengers   6     9       10        13   17
No. of calls                5
```

```
                9  5│9  4│5         0│4  8  3  4  3       3  7  5  2  0
No. of passengers  6     6        3
Total passengers   6    12       15
No. of calls              3
```

```
                9  3│0  1│5         9│3  6│1  5│0  3  4  1  3
No. of passengers  6     1        4         3    3
Total passengers   6     7       11        14   17
No. of calls                5
```

```
                3  6│3  7│0         1│0  3│5  6│    │0  3│7  6│8
No. of passengers  3     3        1         1    4        1    4
Total passengers   3     6        7         8   12       13   17
No. of calls                7
```

(Note: It is alright to create a two-digit number by using adjacent ends of five-digit groups, as in "54" in the first row of the simulation.)

(5) The numbers of parents that called before we had the required passenger space were 3, 5, 5, 5, and 7. The mean is $\bar{x} = 5$ parents.

Linear Transformations of Random Variables

It is sometimes necessary to change or transform a random variable X by adding (or subtracting) a constant, multiplying (or dividing) by a constant, or both. Doing so affects measures of center and spread in a predicable manner.

If a random variable X is transformed by multiplying by the constant b, and/or adding the constant a, then

the mean of the transformed variable is $\mu_{a+bX} = a + b\mu_X$;

the variance of the transformed variable is $\sigma^2_{a+bX} = b^2 \sigma^2_X$.

EXAMPLE: Boxes of a cereal contain collector cards of characters from a movie. Some cards are more common than others. The distribution of the character cards in the boxes is given by the following:

Alfred the Agile: 20%

Brian the Brave: 30%

Craig the Crafty: 50%

The mean number of cereal boxes one would have to buy in order to get all three cards is approximately 6.7 with a standard deviation of 4.1. This was determined by running the earlier simulation 10,000 times! (On a computer, of course.)

Boxes of cereal currently cost $3.72 each. The cereal company also sells a special frame to hold the collectors cards for $5.99. What are the mean and standard deviation of the cost to complete the collection of three cards and put them in the special frame?

ANSWER: The random variable X is the number of boxes purchased to complete the three-card collection. It is given that $\mu X = 6.7$ boxes and $\sigma X = 4.1$ boxes. The variance of the number of boxes purchased would be $\sigma^2_X = (\sigma_X)^2 = 4.1^2 = 16.81$.

For every box purchased, $3.72 is paid. When the set is complete, an additional $5.99 is spent regardless of the number of boxes purchased. The mean and standard deviation of the cost of completing the collection are

$$\mu_{a+bX} = a + b\mu_X$$
$$\mu_{5.99+3.72X} = 5.99 + 3.72(6.7)$$
$$= 30.914$$

and

$$\sigma^2_{a+bX} = b^2\sigma^2_X$$
$$\sigma^2_{5.99+3.72X} = 3.72^2(16.81)$$
$$= 232.624$$

The standard deviation is $\sigma_{a+bX} = \sqrt{\sigma^2_{a+bX}} = \sqrt{232.623504} = 15.252$.

Thus, the mean cost of completing the collection is about $30.91 with a standard deviation of $15.25.

Combining Independent Random Variables

Recall that a random variable is a numerical outcome of a random phenomenon. The heights of American male and female adults are random variables. So are math and verbal scores on college entrance exams. Sometimes, a new random variable is desired by combining or finding the difference between these random variables. When sums and differences of random variables are considered, their measures of center behave in predictable ways.

If a random variable X has mean μ_X and a random variable Y has mean μ_Y, then

the mean of the sum of the variables is $\mu_{X+Y} = \mu_X + \mu_Y$;

the mean of the difference of the variables is $\mu_{X-Y} = \mu_X - \mu_Y$.

The mean of the sum is the sum of the means. The mean of the difference is the difference of the means.

EXAMPLE: A particular college entrance exam has two parts: math and verbal. The distribution of math scores M has a mean of 500 and a standard deviation of 100. The distribution of verbal scores V also has a mean of 500 and a standard deviation of 100. What is the average combined score (math + verbal) on the exam?

ANSWER: It is given that $\mu_M = 500$ and that $\mu_V = 500$, so the mean of the combined score is $\mu_{M+V} = \mu_M + \mu_V = 500 + 500 = 1,000$.

EXAMPLE: American adult males have heights that are distributed with a mean of 173 cm and a standard deviation of 7.5 cm. American adult females have heights that are distributed with a mean of 161 cm and a standard deviation of 6.5 cm. If one male and one female are randomly selected from the population, what is the average difference in their heights?

ANSWER: It is given that $\mu_{male} = 173$ cm and that $\mu_{female} = 161$ cm, so the mean difference in heights is $\mu_{male - female} = \mu_{male} - \mu_{female} = 173 - 161 = 12$ cm.

Knowing only the mean of the combined random variables is usually insufficient. A measure of spread is also needed. The variance and standard deviation of a combination of random variables can be determined if the variables themselves are **independent**. Random variables are independent if there is no correlation between them. If variables are not independent—that is, they are correlated in some way—they are called **dependent**.

The college entrance exam referenced earlier has math and verbal scores. Are these random variables independent? No, they are not. It is likely that students with high math scores will tend to have higher verbal scores, and vice versa. That is, verbal and math scores will tend to be positively correlated.

If a male and a female are randomly selected from the adult population, would the heights of the pair be correlated? It is unlikely. It is safe to say that the heights of randomly selected males and females from the population are independent random variables. However, if we compare male and female adult heights by randomly selecting a married *couple*, those heights probably are not independent. Taller men may tend to marry taller women and vice versa.

If a random variable X has variance σ_X^2 and a random variable Y has variance σ_Y^2, then the variance of the sum of the variables is $\sigma_{X+Y}^2 = \sigma_X^2 + \sigma_Y^2$;

the variance of the difference of the variables is $\sigma_{X-Y}^2 = \sigma_X^2 + \sigma_Y^2$.

The variance of the sum is the sum of the variances. The variance of the difference is also the *sum* of the variances.

EXAMPLE: American adult males have heights that are distributed with a mean of 173 cm and a standard deviation of 7.5 cm. American adult females have heights that are distributed with a mean of 161 cm and a standard deviation of 6.5 cm. If one male and one female are randomly selected from the population, what is the standard deviation of the difference in their heights?

ANSWER: It is given that $\sigma_{male} = 7.5$ cm and that $\sigma_{female} = 6.5$ cm, so the variances are $\sigma_{male}^2 = 56.25$ and $\sigma_{female}^2 = 42.25$. The variance of the difference in heights is $\sigma_{male-female}^2 = \sigma_{male}^2 + \sigma_{female}^2 = 56.25 + 42.25 = 98.5$ cm². The standard deviation of the difference is therefore $\sigma_{male-female} = \sqrt{98.5} \approx 9.9$ cm.

EXAMPLE: A particular college entrance exam has two parts: math and verbal. The distribution of math scores M has a mean of 500 and a standard deviation of 100. The distribution of verbal scores V also has a mean of 500 and a standard deviation of 100. What is the standard deviation of the combined score (math + verbal) on the exam?

ANSWER: Since the random variables, math score and verbal score, are not independent, one cannot calculate the standard deviation of the combined score.

(Note: Actually one can calculate it, but it requires information about the correlation between the variables. It is not a topic on the AP Statistics exam.)

The Normal Distribution

Probability Distributions of Continuous Random Variables

Probability distributions of discrete random variables consist of a function that matches all possible outcomes of a random phenomenon with their associated probabilities. Continuous random variables have probability distributions as well, but because there are an infinite number of outcomes, a definite probability cannot be matched with any particular outcome.

Probability distributions for continuous random variables are usually defined with a mathematical formula or **density curve**. Probabilities are determined for ranges of outcomes by finding the area under the curve. Probability distributions defined for continuous random variables using density curves must meet the same criteria as discrete random variables. The key is that the sum of all probabilities must be 1. With continuous random variables the total area under the curve is 1.

The probability that the continuous random variable X lies between any two values is equal to the area under the curve between the two values. In the two continuous probability density functions below, $P(5 \leq X \leq 10)$ equals 0.5 and 0.75, respectively.

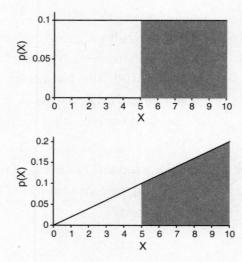

Properties of the Normal Distribution

The **normal distribution** is a continuous probability distribution that appears in many situations, both natural and man-made. It is also, as will be described later, the basis for many of the statistical inference procedures covered by the AP Statistics curriculum. The normal distribution has several important properties:

(1) It is perfectly symmetric, unimodal, and bell-shaped. In fact, it is where the term "bell curve" comes from.

(2) The curve continues infinitely in both directions, and is asymptotic to the horizontal axis as it approaches $\pm\infty$.

(3) It is defined by only two parameters: mean μ and standard deviation σ.

(4) The distribution is centered at the mean μ.

(5) The "points of inflection," where the curve changes from curving downward to curving upward, occur at exactly $\pm 1\sigma$.

(6) The total area under the curve equals 1.

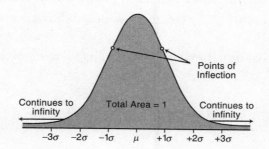

Probabilities of outcomes from random phenomena that have a normal distribution are computed by finding the area under the curve. These areas are determined by first determining how many standard deviations a point lies from the mean. Regardless of the value of the mean and standard deviation, all normal distributions have the same area between the given points. The **68-95-99.7 rule** (sometimes called the **empirical rule**) gives some benchmarks for understanding how probability is distributed under a normal curve.

If a continuous random variable has a normal distribution, approximately 68% of all outcomes will be within 1 standard deviation of the mean. About 95% of all outcomes are within 2 standard deviations of the mean and 99.7% are within 3 standard deviations of the mean. Therefore, only 0.3%—about one outcome in 330—will be more than 3 standard deviations from the mean.

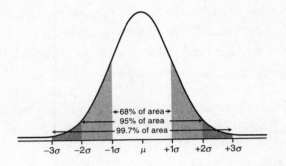

TEST TIP

It's okay to skip an item that you're unsure about or would like to come back to later. However, be careful to also skip that line on your answer sheet so that you continue to fill in your answers on the correct line.

Many phenomena are normally distributed, or at least are very close to being so. For instance, American adult males have heights that are approximately normally distributed with a mean of $\mu = 173$ cm and a standard deviation of $\sigma = 7.5$ cm.

The probability distribution for males would look like this:

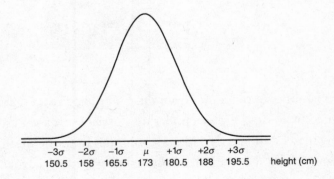

If an adult male were randomly selected from the population, it would not be surprising to find one of height 173 cm or even 180 cm. However, finding a male with a height of 188 cm or more looks unlikely.

EXAMPLE: American adult males have heights that are approximately distributed with a mean of $\mu = 173$ cm and a standard deviation of $\sigma = 7.5$ cm. What percent of males have heights between 165.5 cm and 180.5 cm?

ANSWER: Using the 68-95-99.7 rule, about 68% of all outcomes lie within 1 standard deviation of the mean. Since 165.5 cm is 1 standard deviation below the mean of 173 cm and 180.5 cm is 1 standard deviation above, about 68% of all males have heights between 165.5 cm and 180.5 cm.

EXAMPLE: American adult males have heights that are approximately distributed with a mean of $\mu = 173$ cm and a standard deviation of $\sigma = 7.5$ cm. What percent of males have heights over 188 cm?

ANSWER: Using the 68-95-99.7 rule, about 95% of all outcomes lie within 2 standard deviations of the mean. This means that about 5% fall below 2 standard deviations and above 2 standard deviations combined. Thus, about 2.5% would be above 2 standard deviations from the mean. The height 188 cm is 2 standard deviations above the mean of 173 cm, so roughly 2.5% of all males have heights above 188 cm.

EXAMPLE: A particular college entrance exam has two parts: math and verbal. The distribution of math scores is normal with a mean of 500 and a standard deviation of 100. Between which two values will the middle 99.7% of all test scores roughly lie?

ANSWER: Using the 68-95-99.7 rule, about 99.7% of all outcomes lie within 3 standard deviations of the mean. The standard deviation of the distribution is 100, so 3 standard deviations is 300. The middle 99.7% of all scores will lie within 300 of the mean, or between 200 and 800.

The 68-95-99.7 rule only gives benchmark values for the area under the normal distribution at whole numbers of standard deviations from the mean. If locations other than whole standard deviations are desired, then **Table A: Standard Normal Probabilities** is required.

Determining probabilities from Table A requires one to compute a **standard** score or **z-score**. The z-score is simply the number of standard deviations an observation lies from the mean. The z-score is computed by $z = \dfrac{x - \mu}{\sigma}$, where x is the value of the observed outcome.

EXAMPLE: American adult males have heights that are approximately normally distributed with a mean of $\mu = 173$ cm and a standard deviation of $\sigma = 7.5$ cm. What is the z-score for a male of height 183 cm?

ANSWER: The z-score for a male of height 183 cm is

$$z = \frac{x - \mu}{\sigma} = \frac{183 - 173}{7.5} = \frac{10}{7.5} \approx 1.33.$$

The table of standard normal probabilities gives the probability of observing an outcome below the given z-score. The table is read by cross-referencing the row of the table containing the whole number and tenths place of the z-score with the column holding the hundredths place.

EXAMPLE: What is the value of the standard normal probability table (Table A) for a z-score of 1.33?

ANSWER: Cross-reference the 1.3 row of the table with the 0.03 column.

z	0.00	0.01	0.02	0.03	0.04	...
0.0	0.5000	0.5040	0.5080	0.5120	0.5199	...
.	.	.	.	.	.	...
.	.	.	.	.	.	...
.	.	.	.	.	.	...
1.1	0.8643	0.8665	0.8686	0.8708	0.8729	...
1.2	0.8849	0.8869	0.8888	0.8907	0.8925	...
1.3	0.9032	0.9049	0.9066	**0.9082**	0.9099	...
1.4	0.9192	0.9207	0.9222	0.9236	0.9251	...
1.5	0.9332	0.9345	0.9357	0.9370	0.9382	...
.	.	.	.	.	.	...
.	.	.	.	.	.	...
.	.	.	.	.	.	...

The value of the standard normal probability table (Table A) for a z-score of 1.33 is 0.9082.

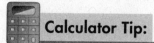 **Calculator Tip:**

Cumulative normal probabilities can be calculated with the graphing calculator. Consider the previous example.

From the home screen, press $\boxed{\text{DISTR}}$, and then choose 2:normalcdf(. This is the normal cumulative density function command and is used for computing normal probabilities (areas under the normal curve) between two values.

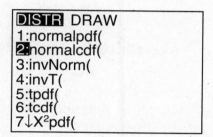

The syntax for the normalcdf command is normalcdf(left bound, right bound, mean, standard deviation). For standard normal distributions (mean = 0 and standard deviation = 1), no mean or standard deviation needs to be entered—the calculator defaults to the values 0 and 1, respectively.

The normal distribution continues to infinity in both the positive and negative directions. When the area in a tail is required and one of the bounds is at infinity, use of an extreme value is needed. In most cases, $-1,000$ or $1,000$ will suffice.

Key in $-1000,1.33)$ and then press $\boxed{\text{ENTER}}$.

```
normalcdf( -1000,
1.33)
         .9082408019
```

The area under the normal curve below $z = 1.33$ is about 0.9082.

EXAMPLE: American adult males have heights that are approximately normally distributed with a mean of $\mu = 173$ cm and a standard deviation of $\sigma = 7.5$ cm. What percent of adult males have heights less than 183 cm?

ANSWER: Since Table A gives the gives the probability of observing an outcome below the given z-score, and the z-score for 183 cm is $z = 1.33$, about 90.82% of adult males have heights less than 183 cm.

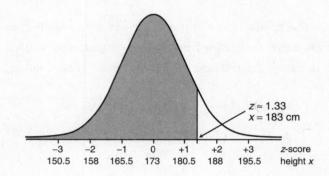

Note: Using probability notation, the above solution would be written as $P(x < 183 \text{ cm}) \approx P(z < 1.33) \approx 0.9082$.

Calculator Tip:

From the home screen, press DISTR, and then choose 2:normalcdf(.

The syntax for the normalcdf command is normalcdf (left bound, right bound, mean, standard deviation). For nonstandard normal distributions (those not having mean = 0 and standard deviation = 1), mean and standard deviation need to be entered.

Key in −1000,183,173,7.5) and then press ENTER.

```
normalcdf( −1000,
183, 173, 7.5)
       .9087887181
■
```

The area under the normal curve, with mean = 173 and standard deviation = 7.5, below $x = 183$ is about 0.9088.

Note: The difference between the previous example's area of 0.9082 and the current area of 0.9088 is due to the fact that the z-score used earlier is not exactly 1.33. A rounded value was used for z, which was, in fact, a little higher than 1.33.

Note: Because a definite probability cannot be matched with any particular outcome, only a range of outcomes, the chance of observing a male with a height of less than 183 cm is the same as the chance of observing a male with a height of 183 cm or less. That is $P(x < 183 \text{ cm}) = P(x \le 183 \text{ cm})$.

EXAMPLE: A particular college entrance exam has two parts: math and verbal. The distribution of math scores is normal with a mean of 500 and a standard deviation of 100. An advanced summer math program requires a score of at least 625 to participate. What proportion of students is eligible to participate in the program?

ANSWER: First, the z-score must be computed. $z = \dfrac{x - \mu}{\sigma} = \dfrac{625 - 500}{100} = 1.25$.

Cross-indexing row 1.2 with column 0.05 in Table A, the result is 0.8944. The table gives the proportion below a score of 625, so the proportion of students with scores of at least 625 is $1 - 0.8944 = 0.1056$.

EXAMPLE: American adult females have heights that are normally distributed with a mean of 161 cm and a standard deviation of 6.5 cm. A company is seeking women of a certain height to operate a particular vehicle. Women should not be shorter than 146 cm (because they will not be able to reach the controls) or taller than 173 cm (because they hit their heads on the roof). What percent of women could hold this job?

ANSWER: The z-scores for 146 cm and 173 cm are $z = \dfrac{146 - 161}{6.5} \approx -2.31$ and $z = \dfrac{173 - 161}{6.5} \approx 1.85$, respectively. Table A shows that 0.0104 of females are below 146 cm and 0.9678 of females are below 173 cm. Therefore, the proportion of females between 146 cm and 173 cm is $0.9678 - 0.0104 = 0.9574$. Almost 96% of women could operate this vehicle.

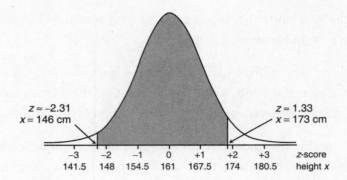

Calculator Tip:

From the home screen, press ⌈DISTR⌋, and then choose 2:normalcdf(. Key in 146,173,161,6.5) and then press ⌈ENTER⌋.

> normalcdf(146, 17
> 3, 161, 6.5)
>   .9570570251

The area under the normal curve, with mean = 161 and standard deviation = 6.5, between $x = 146$ and $x = 173$ is about 0.9571.

Note: The difference between the table-calculated area of 0.9574 and the calculator result of 0.9571 is, again, due to the round-off error in the by-hand calculations.

It is possible to determine outcomes (or mean or standard deviation) given a probability. To do this, one first has to read Table A in "reverse." That is, find the probability in the center of the table and read horizontally and vertically to the margins.

EXAMPLE: What *z*-score corresponds to a Table A value of 0.8907?

z	0.00	0.01	0.02	**0.03**	0.04	...
0.0	0.5000	0.5040	0.5080	0.5120	0.5199	...
.	.	.	.	.	.	...
.	.	.	.	.	.	...
.	.	.	.	.	.	...
1.1	0.8643	0.8665	0.8686	0.8708	0.8729	...
1.2	0.8849	0.8869	0.8888	0.8907	0.8925	...
1.3	0.9032	0.9049	0.9066	0.9082	0.9099	...
1.4	0.9192	0.9207	0.9222	0.9236	0.9251	...
1.5	0.9332	0.9345	0.9357	0.9370	0.9382	...
.	.	.	.	.	.	...
.	.	.	.	.	.	...
.	.	.	.	.	.	...

ANSWER: Since the probability is greater than 0.5, the *z*-score must be positive. Locating 0.8869 and reading "backward" gives a *z*-value of 1.23.

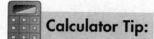

Calculator Tip:

Inverse cumulative normal probabilities can be calculated with the graphing calculator. Consider the previous example.

From the home screen, press ⬚DISTR⬚, and then choose **3:invNorm(**. This is the inverse normal cumulative density function command and is used for finding *z*-scores or nonstandard normal scores given the area in the left tail of a normal distribution.

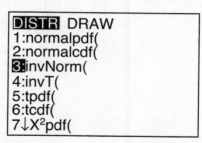

```
DISTR DRAW
1:normalpdf(
2:normalcdf(
3:invNorm(
4:invT(
5:tpdf(
6:tcdf(
7↓X²pdf(
```

The syntax for the invNorm command is invNorm(area, mean, standard deviation). For standard normal distributions (mean = 0 and standard deviation = 1), no mean or standard deviation need to be entered—the calculator defaults to the values 0 and 1, respectively.

Key in .89) and then press ENTER.

The z-score that corresponds to an area under the normal curve in the left tail of 0.89 is about $z = 1.23$.

Note: The calculator can use the value 0.89 rather than the closest value of 0.8907 from the table.

Once a z-score is determined, the unknown values in the equation $z = \dfrac{x - \mu}{\sigma}$ can be determined.

EXAMPLE: A particular college entrance exam has two parts: math and verbal. The distribution of verbal scores is normal with a mean of 500 and a standard deviation of 100. What is the verbal score of a student who scores in the 89th percentile?

ANSWER: The 89th percentile is the point where 89% of outcomes are at or below that point. The probability in the table that is closest to 0.89 is 0.8907, which corresponds to a z-score of 1.23. Solving $z = \dfrac{x - \mu}{\sigma}$ for the unknown x, $x = \mu + z\sigma = 500 + 1.23(100) = 623$. Thus, a score of 623 corresponds to the 89th percentile.

Calculator Tip:

From the home screen, press DISTR , and then choose 3:invNorm(.

The syntax for the invNorm command is invNorm(area, mean, standard deviation). For nonstandard normal distributions (those not having mean = 0 and standard deviation = 1), mean and standard deviation need to be entered.

Key in .89,500,100) and then press ENTER .

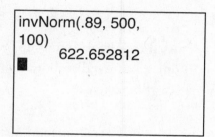

```
invNorm(.89, 500,
100)
        622.652812
```

The test score that corresponds to an area under the normal curve in the left tail of 0.89, with mean 500 and standard deviation 100, is about $x = 622.6$.

(Note: The difference between the hand-and-table-calculated value of 623 and the calculator result of 622.6 is due to using the closest value in the table to the 89th percentile, 0.8907. The calculator uses 0.89.)

EXAMPLE: American adult males have heights that are approximately normally distributed with a mean of $\mu = 173$ cm and a standard deviation of $\sigma = 7.5$ cm. Between which two values are the middle 90% of all male heights?

ANSWER: Considering the middle 90% of males leaves 10% of the males in the two tails of the distribution, or 5% in each distribution. Finding the area in Table A closest to 0.0500 and reading its corresponding z-score gives $z = -1.645$. This is the z-score that cuts off the lowest 5% of the distribution. By symmetry, the top 5% will be cut off by $z = 1.645$.

Solving $-1.645 = \dfrac{x - 173}{7.5}$ and $1.645 = \dfrac{x - 173}{7.5}$ for x, gives values of $x = 160.7$ cm and $x = 185.3$ cm, respectively. The middle 90% of male heights are between these two values.

Normal Approximation of the Binomial Distribution

Under certain conditions, a binomial distribution can be approximated by a normal distribution. The key word is *approximated*, as the binomial distribution is a particular discrete random variable and the normal distribution models continuous random variables.

For a binomial distribution with parameters n and p, it is well approximated by a normal distribution if $np \geq 10$ and $n(1 - p) \geq 10$. Another way to look at this is that np is the mean number of successes and $n(1 - p)$ is the mean number of failures.

EXAMPLE: A tossed tack lands "point up" with a probability of 0.42. If it is tossed 50 times, can the distribution of successes be approximated with a normal distribution?

ANSWER: The situation stated is binomial with $n = 50$ and $p = 0.42$, so $np = (50)(0.42) = 21$ and $n(1 - p) = (50)(0.58) = 29$. Yes, this binomial situation can be approximated with a normal distribution, since $np \geq 10$ and $n(1 - p) \geq 10$.

Recall that the mean of a binomial distribution is $\mu_X = np$ and the standard deviation is $\sigma_X = \sqrt{np(1 - p)}$.

EXAMPLE: A tossed tack lands "point up" with a probability of 0.42. If it is tossed 50 times, what is the probability that it will land "point up" at least 25 times?

ANSWER: The mean number of tacks landing "point up" is

$\mu_X = np = 50\,(0.42) = 21$ and the standard deviation is

$\sigma_X = \sqrt{np(1 - p)} = \sqrt{50(0.42)(0.58)} \approx 3.49$. The z-score for 25 successes

is $z = \dfrac{x - \mu}{\sigma} = \dfrac{25 - 21}{3.49} \approx 1.15$. The value of Table A that corresponds

to $z = 1.15$ is 0.8749. Thus, the probability of at least 25 successes is $1 - 0.8749 = 0.1251$.

Sampling Distributions

A **sampling distribution** is the probability distribution of a sample statistic when a sample is drawn from a population. Recall that a probability distribution is all possible outcomes of a random variable and their associated probabilities. A sample statistic is also a random variable. Thus, a sampling distribution is simply a function describing all possible statistics and their corresponding probabilities from all possible samples of a given size.

Sample statistics are random variables. In most cases, the mean and standard deviation of the sampling distribution of the statistic can be calculated from the characteristics of the population. The mean of a sample statistic is the mean of the statistics from all possible samples of a given size. The standard deviation of a sample statistic is the standard deviation of the statistics from all possible samples of a given size.

TEST TIP

You may only work on each section of the test during the stated time period. Looking ahead at the next sections or trying to go back to check your multiple-choice answers when you're supposed to be working on something else can result in having your test scores canceled.

The Sampling Distribution of a Sample Proportion

The **sampling distribution of a sample proportion** $\hat{p}$ is described when all possible simple random samples of size n are taken from a population where the probability of an individual success is p. The value of the sample proportion $\hat{p}$ is $\hat{p} = \dfrac{X}{n} = \dfrac{\text{number of } \textit{successes} \text{ in the sample}}{n}$. It has the following characteristics:

the mean of the sample proportion $\hat{p}$ is $\mu_{\hat{p}} = p$;

the standard deviation of the sample proportion $\hat{p}$ is $\sigma_{\hat{p}} = \sqrt{\dfrac{p(1-p)}{n}}$.

EXAMPLE: In a large high school of 2,500 students, 21% of them are seniors. A simple random sample of 150 students is taken and the proportion of seniors calculated. What are the mean and standard deviation of the sample proportion, $\hat{p}$?

ANSWER: In this case, the population proportion is $p = 0.21$ and the sample size is $n = 150$. The mean of the sample proportion $\hat{p}$ is $\mu_{\hat{p}} = p = 0.21$ and the standard deviation is $\sigma_{\hat{p}} = \sqrt{\dfrac{p(1-p)}{n}} = \sqrt{\dfrac{0.21(0.79)}{36}} \approx 0.033$.

Note: The value of $\sigma_{\hat{p}}$ assumes that the probability of a success remains constant throughout the sampling process. When sampling from a population without replacement, this is not the case, but the formula is reasonably accurate when the sample size is no more than 10% of the population size. In the previous example, the sample size of 150 is clearly less than 10% of 2,500.

EXAMPLE: A tossed tack lands "point up" with probability 0.42. If the tack is tossed 50 times, what are the mean and standard deviation of $\hat{p}$, the proportion of times the tack lands "point up"?

ANSWER: The probability of landing "point up" is 0.42; this is considered the population proportion $p = 0.42$. The number of tosses is the sample size $n = 50$. The mean of the sample proportion $\hat{p}$ is $\mu_{\hat{p}} = p = 0.42$ and the standard deviation is $\sigma_{\hat{p}} = \sqrt{\dfrac{p(1-p)}{n}} = \sqrt{\dfrac{0.42(0.58)}{50}} \approx 0.070$.

The distribution of sample proportions is closely related to the binomial distribution. In fact, the mean and standard deviation of the sample proportion $\hat{p}$ are simply the mean and standard deviation of the binomial random variable X when divided by the sample size n. Because of this, the sampling distribution of sample proportions can be modeled with a normal distribution like the binomial distribution if $np \geq 10$ and $n(1 - p) \geq 10$.

EXAMPLE: In a large high school of 2,500 students, 21% of them are seniors. A simple random sample of 150 students is taken and the proportion of seniors calculated. What is the probability that the sample will contain less than 15% seniors?

ANSWER: Since the binomial conditions $np \geq 10$ and $n(1 - p) \geq 10$ are met—$np = 150(0.21) = 31.5$ and $n(1 - p) = 150(0.79) = 118.5$—the sampling distribution of $\hat{p}$ can be approximated by a normal distribution with mean $\mu_{\hat{p}} = 0.21$ and standard deviation $\sigma_{\hat{p}} \approx 0.033$.

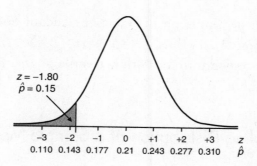

The z-score for 15% is $z = \dfrac{0.15 - 0.21}{\sqrt{\dfrac{0.21(0.79)}{150}}} \approx -1.80$. From Table A, the probability of

z being less than -1.80 is 0.0359. Thus, the chance of obtaining a sample with less than 15% seniors is about 3.6%—not very likely.

EXAMPLE: A tossed tack lands "point up" with probability 0.42. If the tack is tossed 50 times, what proportions of successes correspond to the highest 10% of all possible outcomes?

ANSWER: From the previous example, the mean of the sample proportion $\hat{p}$ is $\mu_{\hat{p}} = 0.42$ and the standard deviation is $\sigma_{\hat{p}} \approx 0.070$. Because np and $n(1 - p)$ are both at least 10—21 and 29, respectively—we can approximate the distribution of $\hat{p}$ with a normal distribution.

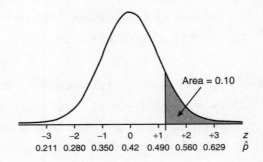

The 90th percentile on the table of standard normal probabilities corresponds to a z-score of about 1.28. Thus, $\hat{p} = \mu + z\sigma = 0.42 + 1.28(0.070) \approx 0.51$. A success rate of $\hat{p} = 0.51$ corresponds to the 90th percentile, so the highest 10% of all possible outcomes are 51% or more.

The Sampling Distribution of a Sample Mean

The **sampling distribution of a sample mean** $\bar{x}$ is described when all possible simple random samples of size n are taken from a population with mean μ and standard deviation σ. It has the following characteristics:

the mean of the sample mean $\bar{x}$ is $\mu_{\bar{x}} = \mu$.

the standard deviation of the sample mean is $\bar{x}$ is $\sigma_{\bar{x}} = \dfrac{\sigma}{\sqrt{n}}$.

EXAMPLE: In a large high school of 2,500 students, the mean number of cars owned by students' families is 2.35 with a standard deviation of 1.06. A simple random sample of 36 students is taken and the mean number of cars owned is calculated. What are the mean and standard deviation of the sample mean, $\bar{x}$?

ANSWER: The population mean is $\mu = 2.35$ cars, the standard deviation is $\sigma = 1.06$ cars, and the sample size is $n = 36$. The mean of the sample mean $\bar{x}$ is $\mu_{\bar{x}} = \mu = 2.35$ cars and the standard deviation is

$$\sigma_{\bar{x}} = \frac{\sigma}{\sqrt{n}} = \frac{1.06}{\sqrt{36}} \approx 0.177 \text{ cars.}$$

Note: The value of $\sigma_{\bar{x}}$ assumes sampling with replacement, which is unlikely in practice. When sampling from a population without replacement, the formula for $\sigma_{\bar{x}}$ is reasonably accurate when the sample size is no more than 10% of the population size, or data come from a randomized experiment. In the previous example, the sample size of 36 is clearly less than 10% of 2,500.

The shape of the sampling distribution of the sample mean $\bar{x}$ is dependent on the shape of the original population. If the population is normally distributed, the sampling distribution of $\bar{x}$ will always be normally distributed. If the population distribution is skewed, the sampling distribution of $\bar{x}$ will also *tend* to be skewed in the same way. However, as the sample size n increases, the shape of the distribution of $\bar{x}$ looks less like the population and more like a normal distribution. This characteristic of the sampling distribution of $\bar{x}$ is so important that it has a special name: the **central limit theorem**.

Central limit theorem: As the size n of a simple random sample increases, the shape of the sampling distribution of $\bar{x}$ tends toward being normally distributed.

The consequence of the central limit theorem is that if the sample size n is sufficiently "large," the sampling distribution of $\bar{x}$ can be approximated by a normal distribution.

EXAMPLE: In a large high school of 2,500 students, the mean number of cars owned by students' families is 2.35 with a standard deviation of 1.06. A histogram of the population is shown.

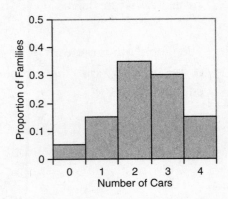

A simple random sample of 36 students is taken and the mean number of cars owned is calculated. What is the probability that the sample mean is greater than 2.5 cars?

ANSWER: The population mean is $\mu = 2.35$ cars, the standard deviation is $\sigma = 1.06$ cars, and the sample size is $n = 36$. The mean of the sample mean $\bar{x}$ is $\mu_{\bar{x}} = \mu = 2.35$ cars and the standard deviation is $\sigma_{\bar{x}} = \dfrac{\sigma}{\sqrt{n}} = \dfrac{1.06}{\sqrt{36}} \approx 0.177$ cars. Since the population distribution is fairly symmetric, a sample size of 36 should be sufficiently large enough to allow the sampling distribution of $\bar{x}$ to be approximated by a normal distribution.

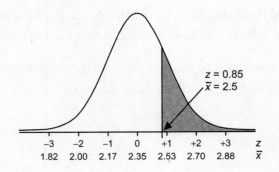

The z-score for a sample mean of 2.5 cars is $z = \dfrac{2.5 - 2.35}{1.06 / \sqrt{36}} \approx 0.85$. From Table A, the probability of z being less than 0.85 is 0.8023. The chance of obtaining a sample with a mean of more than 2.5 cars is $1 - 0.8023 = 0.1977$, or about 19%.

How "large" is large enough? That depends on how far from normal the population distribution is. A rule of thumb is that a little skewness can be tolerated for samples up to about size 15, moderate skewness up to about size 40, and heavy skewness for samples over size 40.

As with proportions, one may also determine the value of a particular percentile rank in a sampling distribution of the sample mean.

EXAMPLE: In a large high school of 2,500 students, the mean number of cars owned by students' families is 2.35 with a standard deviation of 1.06. A simple random sample of 36 students is taken. Below what value is the lowest 5% of all possible sample means?

ANSWER: From the previous examples, the mean of the sample mean $\bar{x}$ is $\mu_{\bar{x}} = \mu = 2.35$ cars and the standard deviation is

$$\sigma_{\bar{x}} = \frac{\sigma}{\sqrt{n}} = \frac{1.06}{\sqrt{36}} \approx 0.177 \text{ cars.}$$

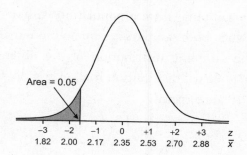

The 5th percentile in Table A corresponds to a z-score of -1.65. Therefore, $\bar{x} = \mu + z\sigma = 2.35 + (-1.65)(0.177) \approx 2.058$. The lowest 5% of all sample means will be below 2.058 cars.

The Sampling Distribution of the Difference Between Two Independent Sample Proportions

The **sampling distribution of the difference between two independent sample proportions** $\hat{p}_1 - \hat{p}_2$ is described when all possible simple random samples of size n_1 are taken from one population and of size n_2 from a second population, and the probabilities of an individual success in each population are p_1 and p_2. The samples must be independent, that is, there should not be a correlation between individuals of the two populations.

The value of the difference in sample proportions $\hat{p}_1 - \hat{p}_2$ is $\hat{p}_1 - \hat{p}_2 = \dfrac{X_1}{n_1} - \dfrac{X_2}{n_2}$. It has the following characteristics:

the mean of the difference between sample proportions $\hat{p}_1 - \hat{p}_2$ is $\mu_{\hat{p}_1 - \hat{p}_2} = p_1 - p_2$;

the standard deviation of the difference between sample proportions $\hat{p}_1 - \hat{p}_2$ is

$$\sigma_{\hat{p}_1 - \hat{p}_2} = \sqrt{\dfrac{p_1(1 - p_1)}{n_1} + \dfrac{p_2(1 - p_2)}{n_2}} \; .$$

As with a single sample for a proportion, the value of $\sigma_{\hat{p}_1 - \hat{p}_2}$ assumes that the probabilities of a success in each population remain constant throughout the sampling process. When sampling from a population without replacement the formula is reasonably accurate when both sample sizes are no more than 10% of their respective population sizes. Also, the sampling distribution of the difference in sample proportions can be modeled with a normal distribution if all of $n_1 p_1 \geq 10$, $n_1(1 - p_1) \geq 10$, $n_2 p_2 \geq 10$, and $n_2(1 - p_2) \geq 10$ are satisfied.

EXAMPLE: In a large high school with 525 seniors, 23% of them own a car. A second high school has 510 seniors, 21% of whom own cars. Simple random samples of 50 seniors are taken from each school, and the proportion of seniors owning cars in each is calculated. What is the likelihood that the survey will indicate that more seniors in the second school own cars than the first school?

ANSWER: Since both samples are no more than 10% of the population, the formula for the standard deviation of $\hat{p}_1 - \hat{p}_2$ will work. Since $n_1 p_1 = (50)(0.23) = 11.5$, $n_1(1 - p_1) = (50)(0.77) = 38.5$, $n_2 p_2 = (50)(0.21) = 10.5$, and $n2(1 - p2) = (50)(0.79) = 39.5$ are all at least 10, the difference $\hat{p}_1 - \hat{p}_2$ can be approximated by the normal distribution.

The mean of the difference in sample proportions $\hat{p}_1 - \hat{p}_2$ is $\mu_{\hat{p}_1 - \hat{p}_2} = p_1 - p_2 = 0.23 - 0.21 = 0.02$ and the standard deviation is $\sigma_{\hat{p}_1 - \hat{p}_2} = \sqrt{\dfrac{0.23(0.77)}{50} + \dfrac{0.21(0.79)}{50}} \approx 0.083$.

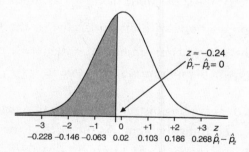

The question asks the probability that $\hat{p}_1 < \hat{p}_2$ or $\hat{p}_1 - \hat{p}_2 < 0$. The value of $\hat{p}_1 - \hat{p}_2 = 0$ has a z-score of $z = \dfrac{0 - 0.02}{0.083} \approx -0.24$. From Table A, the probability of a z-score being -0.24 or less is 0.4052. Thus, the chance that the survey would show a greater proportion of seniors owning cars in the second school than the first is about 41%, a very likely occurrence.

DIDYOUKNOW?

The best-selling car in the United States is actually a pickup truck. The Ford F-Series of trucks has been the top-selling vehicle in the United States for over 30 years.

The Sampling Distribution of the Difference Between Two Independent Sample Means

The **sampling distribution of the difference between two independent sample means** $\bar{x}_1 - \bar{x}_2$ is described when all possible simple random samples of size n_1 are taken from one population and of size n_2 from a second population, and the parameters of the populations are μ_1, σ_1, μ_2, and σ_2. The samples must be independent, that is, there should not be a correlation between individuals of the two populations. The value of the difference in sample means $\bar{x}_1 - \bar{x}_2$ has the following characteristics:

the mean of the difference between sample means $\bar{x}_1 - \bar{x}_2$ is $\mu_{\bar{x}_1 - \bar{x}_2} = \mu_1 - \mu_2$;

the standard deviation of the difference between sample means $\bar{x}_1 - \bar{x}_2$ is

$$\sigma_{\bar{x}_1 - \bar{x}_2} = \sqrt{\frac{\sigma_1^2}{n_1} + \frac{\sigma_2^2}{n_2}}.$$

As with single samples, the value of $\sigma_{\bar{x}_1 - \bar{x}_2}$ assumes sampling with replacement. When sampling from populations without replacement, the formula for $\sigma_{\bar{x}_1 - \bar{x}_2}$ is reasonably accurate when the sample sizes are no more than 10% of the population size or the data come from a completely randomized experiment.

The shape of the sampling distribution of the difference in sample means $\bar{x}_1 - \bar{x}_2$ is also subject to the central limit theorem. As the sizes of the two independent simple random samples increase, the shape of the sampling distribution of $\bar{x}_1 - \bar{x}_2$ tends toward being normally distributed. Skewness or non-normality in the population distributions is less of a factor with two independent samples and the distribution of $\bar{x}_1 - \bar{x}_2$ can be modeled well by the normal distribution for even small samples.

EXAMPLE: In a large high school of 2,500 students, the mean number of cars owned by students' families is 2.35 with a standard deviation of 1.06. A second large high school has 2,350 students. The mean number of cars owned by students' families is 2.01 with standard deviation of 1.19.

Simple random samples of 36 students are taken from each school, and the mean number of cars per family is calculated for each. What is the likelihood that the survey will indicate that the average number of cars owned by families in the second school is greater than the first?

ANSWER: Since both samples are no more than 10% of the population, the formula for the standard deviation of $\bar{x}_1 - \bar{x}_2$ will work. Unless either or both of the populations are extremely skewed, the sample sizes of 36 are sufficiently large to allow the normal distribution to model the sampling distribution of $\bar{x}_1 - \bar{x}_2$.

The mean of the difference in sample means $\bar{x}_1 - \bar{x}_2$ is $\mu_{\bar{x}_1 - \bar{x}_2} = \mu_1 - \mu_2 =$

$2.35 - 2.01 = 0.34$ cars, and the standard deviation is $\sigma_{\bar{x}_1 - \bar{x}_2} = \sqrt{\dfrac{1.06^2}{36} + \dfrac{1.19^2}{36}} \approx 0.266$ cars.

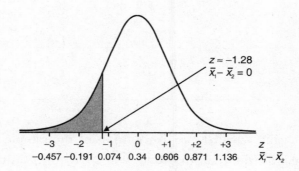

The question asks the probability that $\bar{x}_1 < \bar{x}_2$ or $\bar{x}_1 - \bar{x}_2 < 0$. The value of $\bar{x}_1 - \bar{x}_2 = 0$ has a z-score of $z = \dfrac{0 - 0.34}{0.266} \approx -1.28$. From Table A, the probability of a z-score being -1.28 or less is 0.1003. Thus, the chance that the survey would show a greater average number of cars per family in the second school than the first is about 10%.

Standard Error

It is a rare occasion that one knows the parameter(s) of population(s); after all, part of statistics is determining what those parameters are. When one does not know the parameters necessary to calculate the standard deviation of the sample statistic, then one estimates them with the **standard error**.

In the case of proportions, if p is unknown, the sample proportion $\hat{p}$ will be used to estimate it. In the case of means, if σ is unknown, the sample standard deviation s will be used to estimate it.

Sample Statistic	Standard Deviation	Standard Error
$\hat{p}$	$\sigma_{\hat{p}} = \sqrt{\dfrac{p(1-p)}{n}}$	$SE_{\hat{p}} = \sqrt{\dfrac{\hat{p}(1-\hat{p})}{n}}$
$\bar{x}$	$\sigma_{\bar{x}} = \dfrac{\sigma}{\sqrt{n}}$	$SE_{\bar{x}} = \dfrac{s}{\sqrt{n}}$
$\hat{p}_1 - \hat{p}_2$	$\sigma_{\hat{p}_1 - \hat{p}_2} = \sqrt{\dfrac{p_1(1-p_1)}{n_1} + \dfrac{p_2(1-p_2)}{n_2}}$	$SE_{\hat{p}_1 - \hat{p}_2} = \sqrt{\dfrac{\hat{p}_1(1-p_1)}{n_1} + \dfrac{\hat{p}_2(1-p_2)}{n_2}}$
$\bar{x}_1 - \bar{x}_2$	$\sigma_{\bar{x}_1 - \bar{x}_2} = \sqrt{\dfrac{\sigma_1^2}{n_1} + \dfrac{\sigma_2^2}{n_2}}$	$SE_{\bar{x}_1 - \bar{x}_2} = \sqrt{\dfrac{s_1^2}{n_1} + \dfrac{s_2^2}{n_2}}$

In assessing whether the distribution of the sample proportion $\hat{p}$ can be modeled by a normal distribution, the same rules apply for $\hat{p}$ as they did with p. Both of $n\hat{p}$ and $n(1-\hat{p})$ must be at least 10. For $\hat{p}_1 - \hat{p}_2$, all of $n_1\hat{p}_1$, $n_1(1-\hat{p}_1)$, $n_2\hat{p}_2$, and $n_2(1-\hat{p}_2)$, must be at least 10.

For the sample mean $\bar{x}$ and difference of sample means $\bar{x}_1 - \bar{x}_2$, other issues must be addressed with regard to the shape of the sampling distribution if the population standard deviation(s) is/are unknown.

t-distributions

When the sample standard deviation s is used as an estimate of the population standard deviation σ, the distribution of the standardized score $\dfrac{\bar{x} - \mu}{s/\sqrt{n}}$ is not well approximated by the normal distribution for small and moderate sample sizes. Even if the population itself is normally distributed, the sampling distribution of $\dfrac{\bar{x} - \mu}{s/\sqrt{n}}$ will not be. In this case, a new distribution is required: the **t-distribution**.

The *t*-distribution, sometimes known as Student's *t*, is actually a family of curves that have the following characteristics:

(1) They are symmetric and unimodal, and have a mean equal to zero, like the normal distribution.

(2) The height of the *t*-distribution is shorter at the mean than that of the normal distribution and thicker in the tails.

(3) Each curve in the family is defined by its **degrees of freedom**.

(4) As the number of degrees of freedom increases, the curve approaches a normal distribution.

(5) The standard deviation of the distribution is greater than 1, and approaches 1 as the number of degrees of freedom increases.

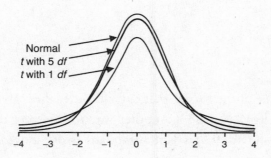

Table B is the table of **t-distribution critical values**. The values in the table give t-critical values, labeled **t***, that cut off a certain percentage of the curve area in the right tail.

The table can be used in three ways. The first is by cross-referencing the row containing the number of degrees of freedom *df* and the probability that a t-score will be above a certain critical value, t^*. The result is that value.

EXAMPLE: What value of the t-score cuts off the highest 10% of a t-distribution with 9 degrees of freedom?

ANSWER: Cross-reference the 9 *df* row of the table with the 0.10 column.

	Tail probability p					
df	0.25	0.20	0.15	0.10	0.05	...
1	1.000	1.376	1.963	3.078	6.314	...
.	.	.	.	.	.	...
.	.	.	.	.	.	...
.	.	.	.	.	.	...
7	0.711	0.896	1.119	1.415	1.895	...
8	0.706	0.889	1.108	1.397	1.860	...
9	0.703	0.883	1.100	**1.383**	1.833	...
10	0.700	0.879	1.093	1.372	1.812	...
11	0.697	0.876	1.088	1.363	1.796	...
.	.	.	.	.	.	...
.	.	.	.	.	.	...
.	.	.	.	.	.	...

The value of the t-score that cuts off the highest 10% of a t-distribution with 9 degrees of freedom is $t^* = 1.383$. Another way to interpret this is that the 90th percentile of a t-distribution with 9 *df* is 1.383 standard errors above the mean.

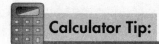
Calculator Tip:

Inverse cumulative t-distribution probabilities can be calculated with the graphing calculator. Consider the previous example.

From the home screen, press $\boxed{\text{DISTR}}$, and then choose 4:invT(. This is the inverse t-distribution cumulative density function command and is used for finding t-scores given the area in the left tail of a t-distribution given the number of degrees of freedom.

```
DISTR DRAW
1:normalpdf(
2:normalcdf(
3:invNorm(
4:invT(
5:tpdf(
6:tcdf(
7↓X²pdf(
```

The syntax for the invT command is invT(area, df).

Key in .9,9) and then press $\boxed{\text{ENTER}}$.

```
invT(.9, 9)
          1.383028738
```

The t-score that corresponds to an area of 0.1 in the right tail (0.9 in the left tail) of the t-distribution with 9 df is about $t = 1.383$.

(Note: Not all calculators come programmed with the invT function.)

Note that Table B contains no negative values for t^*. Since the t-distribution is symmetric, the opposite of t^*, $-t^*$, will cut off an equal area in the left tail as t^*.

EXAMPLE: What value of the t-score cuts off the lowest 5% of a t-distribution with 10 degrees of freedom?

ANSWER: Cross-reference the 10 *df* row of the table with the 0.05 column. The table reads 1.812. Since the area under consideration is the *lowest* 5%, $t^* = -1.812$.

df	Tail probability p					
	0.25	0.20	0.15	0.10	0.05	...
1	1.000	1.376	1.963	3.078	6.314	...
.	.	.	.	.	.	...
.	.	.	.	.	.	...
.	.	.	.	.	.	...
7	0.711	0.896	1.119	1.415	1.895	...
8	0.706	0.889	1.108	1.397	1.860	...
9	0.703	0.883	1.100	1.383	1.833	...
10	0.700	0.879	1.093	1.372	**1.812**	...
11	0.697	0.876	1.088	1.363	1.796	...
.	.	.	.	.	.	...
.	.	.	.	.	.	...
.	.	.	.	.	.	...
	50%	60%	70%	80%	90%	...

Confidence level C

The second way that the table can be used is to find the probability that a t-score will be above a certain critical value, t^*.

EXAMPLE: What is the probability that a *t*-score will be larger than $t^* = 1.100$ with 9 degrees of freedom?

ANSWER: Read across the 9 *df* row of the table until 1.100 is found. Then read the tail probability for that column. The tail probability is 0.15; thus, the probability that a *t*-score will be larger than $t^* = 1.100$ with 9 *df* is 0.15. Another way to interpret this is that a score 1.100 standard errors above the mean, with 9 *df*, is the 85th percentile.

Remember that the *t*-distribution is symmetric and that Table B contains no negative values for t^*. The probability that a t-score will be larger than $t^* = 1.100$ with 9 df is 0.15. Correspondingly, the probability that a *t*-score will be smaller than $t^* = -1.100$ with 9 *df* is also 0.15. This means that with 9 *df*, $t^* = -1.100$ is the 15th percentile.

	Tail probability *p*					
df	0.25	0.20	0.15	0.10	0.05	...
1	1.000	1.376	1.963	3.078	6.314	...
.	.	.	.	.	.	...
.	.	.	.	.	.	...
.	.	.	.	.	.	...
7	0.711	0.896	1.119	1.415	1.895	...
8	0.706	0.889	1.108	1.397	1.860	...
9	0.703	0.883	**1.100**	1.383	1.833	...
10	0.700	0.879	1.093	1.372	1.812	...
11	0.697	0.876	1.088	1.363	1.796	...
.	.	.	.	.	.	...
.	.	.	.	.	.	...
.	.	.	.	.	.	...
	50%	60%	70%	80%	90%	...

Confidence level C

TEST TIP

Be sure to read each question and answer choice carefully. Working too quickly may cause you to miss important key words and answer questions incorrectly.

Calculator Tip:

Cumulative *t*-distribution probabilities can be calculated with the graphing calculator. Consider the previous example.

From the home screen, press DISTR , and then choose **6:tcdf(**. This is the *t*-distribution cumulative density function command and is used for computing *t*-distribution probabilities (areas under the *t*-curve) between two values for a number of degrees of freedom.

```
DISTR DRAW
1:normalpdf(
2:normalcdf(
3:invNorm(
4:invT(
5:tpdf(
6:tcdf(
7↓X²pdf(
```

The protocol for the tcdf command is tcdf(left bound, right bound, df).

The *t*-distribution continues to infinity in both the positive and negative directions. When the area in a tail is required and one of the bounds is at infinity, use of an extreme value is needed. In most cases, −1,000 or 1,000 will suffice.

Key in **1.100,1000,9)** and then press ENTER .

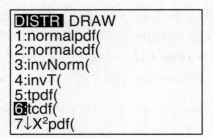

```
tcdf(1.100, 1000,
9)
         .149941381
```

The area under the *t*-curve above $t^* = 1.100$ is about 0.15.

A third way to read the table is to find the values of $\pm t^*$ given the central area of a *t*-distribution, or vice versa.

EXAMPLE: What values of the *t*-score bound the middle 80% of a *t*-distribution with 9 degrees of freedom?

ANSWER: The area under consideration is the middle 80%, so there must be 20% left in the tails. Since the *t*-distribution is symmetric, an equal proportion of the area must exist in each tail. The tail probability is therefore 0.10. Cross-reference the 9 *df* row of the table with the 0.10 column.

df	Tail probability *p*					
	0.25	0.20	0.15	0.10	0.05	...
1	1.000	1.376	1.963	3.078	6.314	...
.	.	.	.	.	.	...
.	.	.	.	.	.	...
.	.	.	.	.	.	...
7	0.711	0.896	1.119	1.415	1.895	...
8	0.706	0.889	1.108	1.397	1.860	...
9	0.703	0.883	1.100	**1.383**	1.833	...
10	0.700	0.879	1.093	1.372	1.812	...
11	0.697	0.876	1.088	1.363	1.796	...
.	.	.	.	.	.	...
.	.	.	.	.	.	...
.	.	.	.	.	.	...
	50%	60%	70%	80%	90%	...

Confidence level *C*

The values of *t*-scores that bound the middle 80% of a *t*-distribution with 9 degrees of freedom are $t^* = \pm 1.383$.

Note that the value at the bottom of the table above the line "Confidence level C" when 9 df and $t^* = 1.383$ are cross-referenced is also 80%. This "confidence level" will be discussed later, but corresponds to central area. Twice the tail probability is equal to the complement of the confidence level, or $1 - 2p = C$.

Calculator Tip:

From the home screen, press $\boxed{\text{DISTR}}$, and then choose 4:invT(.

Key in .1,9) and then press $\boxed{\text{ENTER}}$.

```
invT(.1, 9)
        -1.383028738
```

The t-scores that correspond to a central area of 0.8 (0.1 in the left tail) of a t-distribution with 9 df is about $t = \pm 1.383$.

EXAMPLE: What is the probability that an observation lies within 1.812 standard errors of the mean in a *t*-distribution with 10 degrees of freedom?

ANSWER: Read across the 10 *df* row until 1.812 is found; then read the confidence level (or tail probability) for that column. The confidence level is 90%; thus, the probability of an observation lying within 1.812 standard errors of the mean in the *t*-distribution with 10 *df* is 0.90. (Note: The tail probability is 0.05. The combined probability of both the left and right tails is 0.10, leaving 0.90 in the central area.)

df	Tail probability p					
	0.25	0.20	0.15	0.10	0.05	...
1	1.000	1.376	1.963	3.078	6.314	...
.	.	.	.	.	.	...
.	.	.	.	.	.	...
.	.	.	.	.	.	...
7	0.711	0.896	1.119	1.415	1.895	...
8	0.706	0.889	1.108	1.397	1.860	...
9	0.703	0.883	1.100	1.383	1.833	...
10	0.700	0.879	1.093	1.372	**1.812**	...
11	0.697	0.876	1.088	1.363	1.796	...
.	.	.	.	.	.	...
.	.	.	.	.	.	...
.	.	.	.	.	.	...
	50%	60%	70%	80%	90%	...

Confidence level C

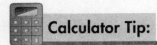 **Calculator Tip:**

From the home screen, press $\boxed{\text{DISTR}}$, and then choose **6:tcdf(**.

Key in -1.812,1.812,10) and then press $\boxed{\text{ENTER}}$.

```
tcdf(−1.812, 1.81
2, 10)
        .8999247413
```

The area under the *t*-curve between $t^* = \pm 1.812$ is about 0.9.

There are times when Table B does not provide exact information. Note that the t^*-values are only for tail probabilities with round numbers: 0.05, 0.025, 0.02, 0.01, etc. It is likely that a computed t^*-value will not be among those in the table. In this case, the table can only provide a range of values for the tail probability *p*, or confidence level (central area) *C*.

EXAMPLE: What is the probability that a *t*-score will be larger than $t^* = 1.5$ with 9 degrees of freedom?

ANSWER: Read across the 9 *df* row of the table until 1.5 is found. The value of 1.5 is not shown in the table, but would be located between 1.383 and 1.833, which correspond to the tail probabilities $p = 0.10$ and $p = 0.05$, respectively. The probability that a *t*-score will be larger than $t^* = 1.5$ with 9 *df* is between 0.05 and 0.10. Thus, $0.05 < p < 0.10$. This is the best information that the table can provide.

		Tail probability p				
df	0.25	0.20	0.15	0.10	0.05	...
1	1.000	1.376	1.963	3.078	6.314	...
.	.	.	.	.	.	...
.	.	.	.	.	.	...
.	.	.	.	.	.	...
7	0.711	0.896	1.119	1.415	1.895	...
8	0.706	0.889	1.108	1.397	1.860	...
9	0.703	0.883	1.100	1.383	**1.833**	...
10	0.700	0.879	1.093	1.372	1.812	...
11	0.697	0.876	1.088	1.363	1.796	...
.	.	.	.	.	.	...
.	.	.	.	.	.	...
.	.	.	.	.	.	...
	50%	60%	70%	80%	90%	...

Confidence level C

To get more precise answers, one would need technology, like a computer or graphing calculator.

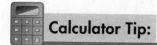

Calculator Tip:

From the home screen, press ⬚DISTR⬚, and then choose 6:tcdf(.

Key in **1.5,1000,9**) and then press ⬚ENTER⬚.

```
tcdf(1.5, 1000, 9)
          .083925328
```

The area under the t-curve above $t^* = 1.5$ is about 0.084.

Table B has lines for only certain numbers of degrees of freedom. There are lines for 30 and 40 df, but not for values between 30 and 40. When the number of degrees of freedom is between two rows in the table, it is recommended that one use the line with the smaller number of degrees of freedom. (For example, for 38 df, use 30.) However, technology will be able to give precise values for any number of degrees of freedom.

Finally, the last line of Table B has the symbol ∞ (infinity). Recall that as the number degrees of freedom increases, the t-curve approaches a normal curve. A t-distribution with an infinite number of degrees of freedom is a normal distribution. Thus, the ∞ line of Table B can be used for normal distributions.

χ^2 (Chi-Square) Distributions

The chi-square distribution is another form of probability distribution, like the normal and t-distributions. The distribution is used to model how far the counts of an observed set of categorical data, with several categories, are from what one would expect given a random sample.

The χ^2-distribution is a family of curves that have the following characteristics:

(1) They are right-skewed, unimodal, and have only non-negative values.

(2) Each curve in the family is defined by its **degrees of freedom, *df***.

(3) As the number degrees of freedom increases, the curve approaches a normal distribution.

(4) The mean of the distribution is equal to the number of degrees of freedom.

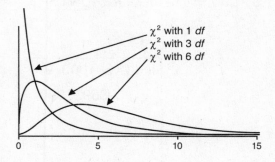

Table C is the table of χ^2 **critical values**. The values in the table give χ^2 critical values that cut off a certain percentage of the curve's area in the right tail.

EXAMPLE: What value of χ^2 cuts off the highest 10% of a χ^2-distribution with 3 degrees of freedom?

ANSWER: Cross-reference the 3 df row of the table with the 0.10 column.

	Tail probability p					
df	0.25	0.20	0.15	0.10	0.05	...
1	1.32	1.64	2.07	2.71	3.84	...
2	2.77	3.22	3.79	4.61	5.99	...
3	4.11	4.64	5.32	**6.25**	7.81	...
4	5.39	5.99	6.74	7.78	9.49	...
5	6.63	7.29	8.12	9.24	11.07	...
.	.	.	.	.	.	...
.	.	.	.	.	.	...
.	.	.	.	.	.	...

The value of χ^2 that cuts off the highest 10% of a χ^2-distribution with 3 degrees of freedom is 6.25. Another way to interpret this is that the 90th percentile of a χ^2-distribution with 3 df is a χ^2-statistic of 6.25.

EXAMPLE: What is the area of the right tail of a χ^2-distribution with 1 degree of freedom cut off by a χ^2-statistic of 2.71?

ANSWER: Read across the 1 df row until 2.71 is found; then read the tail probability for that column. The right-tail probability is 0.10.

	Tail probability p					
df	0.25	0.20	0.15	0.10	0.05	...
1	1.32	1.64	2.07	**2.71**	3.84	...
2	2.77	3.22	3.79	4.61	5.99	...
3	4.11	4.64	5.32	6.25	7.81	...
4	5.39	5.99	6.74	7.78	9.49	...
5	6.63	7.29	8.12	9.24	11.07	...
.	.	.	.	.	.	...
.	.	.	.	.	.	...
.	.	.	.	.	.	...

Cumulative χ^2-distribution probabilities can be calculated with the graphing calculator. Consider the previous example.

From the home screen, press DISTR, and then choose **8: X2cdf(**. This is the χ^2-distribution cumulative density function command and is used for computing χ^2-distribution probabilities (areas under the χ^2-curve) between two values for a number of degrees of freedom.

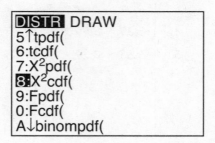

```
DISTR DRAW
5↑tpdf(
6:tcdf(
7:X²pdf(
8:X²cdf(
9:Fpdf(
0:Fcdf(
A↓binompdf(
```

The syntax for the X^2cdf command is X^2cdf(left bound, right bound, df).

The χ^2-distribution continues to infinity in the positive direction. When the area in the tail is required, use of an extreme value is needed. In most cases, 1,000 will suffice.

Key in **2.71,1000,1)** and then press ENTER.

```
X²cdf(2.71, 1000,
1)
       .0997209906
```

The area under the χ^2-curve above $\chi^2 = 2.81$ is about 0.10.

There are times when Table C does not provide exact information. Note that the χ^2-statistics are only for tail probabilities with round numbers: 0.05, 0.025, 0.02, 0.01, etc. It is likely that a computed χ^2-statistic will not be among those in the table. In this case, the table can only provide a range of values for the tail probability p.

EXAMPLE: What is the probability that a χ^2-statistic will be larger than $\chi^2 = 4$ with 2 degrees of freedom?

ANSWER: Read across the 2 *df* row of the table until 4 is found. The value of 4 is not in the table, but would be located between 3.79 and 4.61, which correspond to the tail probabilities $p = 0.15$ and $p = 0.10$, respectively. The probability that a χ^2-statistic will be larger than $\chi^2 = 4$ with 2 *df* is between 0.10 and 0.15. Thus, $0.10 < p < 0.15$. This is the best information that the table can provide.

To get more precise answers, one would need technology, like a computer or graphing calculator.

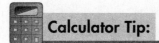

Calculator Tip:

From the home screen, press $\boxed{\text{DISTR}}$, and then choose **8: X2cdf(**.

Key in **4,1000,2**) and then press $\boxed{\text{ENTER}}$.

$$X^2\text{cdf}(4, 1000, 2)$$
$$.1353352832$$

The area under the χ^2-curve above $\chi^2 = 4$ is about 0.135.

The **sampling distribution of the χ^2-statistic** is described when all possible simple random samples of size n are taken from a population where one or two categorical variables, each with two or more categories, are under consideration. The value of the χ^2-statistic is

$$x^2 = \sum_{\text{all cells}} \frac{(\text{observed} - \text{expected})^2}{\text{expected}}.$$

The χ^2-statistic is essentially an aggregate of the relative differences between the observed counts of categories in a sample and the expected counts of those categories based on population characteristics.

Note: The distribution of this χ^2-statistic assumes sampling with replacement, which is unlikely in practice. When sampling from a population without replacement, the χ^2 model is reasonable if the sample size is no more than 10% of the population size. Also, expected counts for all categories should be at least 5.

The number of degrees of freedom for the distribution and the method of computing expected cell counts are dependent on how many populations and how many variables are under study. These issues will be addressed in detail in Chapter 6.

Time for a quiz
- Review strategies in Chapter 2
- Take Quizzes 5 and 6 at the REA Study Center
  (www.rea.com/studycenter)

Chapter 6

Statistical Inference

Estimation (Point Estimators and Confidence Intervals)

Estimation is the process of determining the value of a population **parameter** from information provided by a sample **statistic**. Sometimes, this will be a single value; other times, it will be a range of values.

Point and Interval Estimates

A **point estimate** is a single value that has been calculated to estimate the unknown parameter. The point estimates used in inference are sample statistics. The table below shows which statistics are used as point estimates for parameters.

Parameter	Statistic
Population proportion, p	Sample proportion, $\hat{p}$
Population mean, μ	Sample mean, $\bar{x}$
Difference in population proportions, $p_1 - p_2$	Difference in sample proportions, $\hat{p}_1 - \hat{p}_2$
Difference in population means, $\mu_1 - \mu_2$	Difference in sample means, $\bar{x}_1 - \bar{x}_2$
Population mean difference, μ_d (matched pairs)	Sample mean difference, $\bar{x}_d$
Slope of a population linear model, b_1	Slope of a sample least-squares line, b_1

A **confidence interval** gives us a range of plausible values that is likely to contain the unknown population parameter. This range of values is generated using a set of sample data. The center of a confidence interval for a population parameter is the sample statistic. Confidence intervals are always constructed as *statistic ± margin of error*.

The **margin of error** is the range of values to the left and right of the point estimate in which the parameter likely lies. For example, suppose you read in the school newspaper that 70% of students surveyed said they want more soda machines in the school building. The student who wrote the story reports that the survey has a 10% margin of error. This means it is likely that between 60% and 80% of the entire student body wants more soda machines. We could write this as 70% ± 10%.

Interpreting Confidence Level and Confidence Intervals

We can never be totally certain about what the unknown parameter is, but do have a level of certainty whether or not it is in our interval. This is known as the **confidence level**. The most commonly used confidence levels are 90%, 95%, and 99%. The graph below is an example of a series of 90% confidence intervals created by taking random samples from a population.

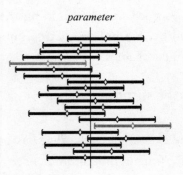

parameter

The vertical line represents the value of the parameter. Each of these intervals shows the form *statistic ± margin of error*. The diamonds in the center of each interval are point estimates of the parameter (statistics), and the bars extend out equal to the margin of error on both sides of the point estimate. Eighteen of the 20 intervals, or 90% of the intervals, intersect the vertical line. This illustrates what 90% confidence means. If we collect repeated samples and construct 90% confidence intervals for the parameter, on average 90% of the intervals generated would contain or "capture" the population parameter.

When we generate a sample and construct a confidence interval, an interpretation of that interval should follow. That interpretation reflects our confidence that the population parameter is within a given range. This is not the same as confidence level.

EXAMPLE: You are interested in finding out what the true proportion of yellow candies is in a very large bag without counting them all. After taking a sample and determining its proportion of yellow, the 95% confidence interval for the true proportion of yellow candies is 0.2 ± 0.1 or $(0.1, 0.3)$. Interpret the interval and describe what is meant by 95% confidence.

ANSWER: I am 95% confident that the true population proportion of yellow candies is between 0.1 and 0.3.

Note: This is not a probability statement about the "chance" that there is a certain proportion of yellow candies in the bag or the chance that the true proportion is actually in our interval. It may be or it may not be—we will never know. It is about how confident we are that our interval has captured the true proportion.

Ninety-five percent confidence means if we collected more samples using the same sampling method and generated confidence intervals, 95% of the intervals would contain the population parameter.

TEST TIP

If you change your answer to a multiple-choice question, be sure to erase your original answer completely. Otherwise, the machines that grade these sections may count your answer wrong because of a double-bubble.

General Formula for Confidence Intervals

Recall that all confidence intervals are in the same format: *statistic $\pm$ margin of error*. The margin of error is the product of two numbers: the **critical value** and the **standard deviation of the statistic**. Thus, a confidence interval can be thought of as *statistic $\pm$ (critical value)(standard deviation of the statistic)*. If the standard deviation of the statistic cannot be computed, then the **standard error** is used. (See Chapter 5, Section D.)

The critical value is found from **Table B: *t*-distribution critical values**, by cross-referencing the confidence level C and the number of degrees of freedom *df*. In cases where the *z*-distribution is used to find the critical value—inference for sample proportions or when the population standard deviation is known in inference for means—the last line (∞) is used.

General Procedure for Inference with Confidence Intervals

To do any inference problem, the four-step procedure for inference is used.

Step 1: State the parameter of interest. The parameter of interest is a characteristic of the population under study. It will be a population proportion, a difference of population means, and so on. This should be defined in words and symbols. (In the next section, *Tests of Significance*, another element will be added to this step.)

Step 2: Name the inference procedure and check assumptions/conditions. The procedure is named by giving the level of confidence and what parameter it is estimating.

There are assumptions/conditions that must be *checked* in order to proceed with various inference procedures. It is not enough to merely state them—you must verify that they apply.

Step 3: Calculate the confidence interval. Confidence intervals are always of the form *statistic ± (critical value)(standard deviation of the statistic)*. Write the appropriate formula for the type of confidence interval being computed, substitute in the values in the formula, and compute the interval.

Step 4: Interpret the results in context. Describe in the context of the situation what the confidence interval means. Refer back to step 1, as this is where the parameter of interest was defined.

Confidence Intervals for Proportions

The formula to compute a confidence interval for a population proportion p is $\hat{p} \pm z^* \sqrt{\dfrac{\hat{p}(1-\hat{p})}{n}}$, where $\hat{p}$ is the sample proportion, n is the sample size, and z^* is the z-critical value.

The assumptions/conditions for inference for a population proportion are as follows:

(1) The sample must be an SRS from the population of interest.

(2) The population is at least 10 times as large as the sample size.

(3) The number of successes and failures are both at least 10. That is, $n\hat{p} \geq 10$ and $n(1-\hat{p}) \geq 10$.

■ **EXAMPLE:** A random sample of 100 candies has 25 that are yellow. Construct the 95% confidence interval for the true proportion of yellow candies.

■ **ANSWER:** Follow the four-step procedure for inference.

Step 1: State the parameter of interest. The parameter of interest is p, the true proportion of yellow candies in the population of all candies.

Step 2: Name the inference procedure and check assumptions/conditions. The procedure is a 95% confidence interval for p, as defined above.

(1) The sample must be an SRS from the population of interest. OK—the problem states that we have a random sample.

(2) The population is at least 10 times as large as the sample size. OK—it is a good bet that there are more than 1,000 candies in the population.

(3) The number of successes and failures are both at least 10. OK—there are 25 yellow candies (successes) and 75 of other colors (failures).

Step 3: Calculate the confidence interval. In our sample $\hat{p} = \dfrac{25}{100} = 0.25$, and the z-critical value for 95% confidence is 1.960. Substituting into the formula $\hat{p} \pm z^* \sqrt{\dfrac{\hat{p}(1-\hat{p})}{n}}$, we get

$$\hat{p} \pm z^* \sqrt{\frac{\hat{p}(1-\hat{p})}{n}}$$

$$0.25 \pm 1.960 \sqrt{\frac{(0.25)(0.75)}{100}}$$

$$0.25 \pm 0.08487$$

$$(0.16513, 0.33487).$$

Step 4: Interpret the results in context. We are 95% confident that the true proportion of yellow candies in the population is between about 0.165 and 0.335.

Calculator Tip:

Confidence intervals for proportions can be calculated with the graphing calculator.

From the home screen, press $\boxed{\text{STAT}}$, arrow right to TESTS, and then choose A:1-PropZInt....

```
EDIT CALC TESTS
8↑ TInterval...
9 : 2-SampZInt...
0 : 2-SampTInt...
A: 1-PropZInt...
B : 2-PropZInt...
C : χ²-Test...
D↓ χ² GOF-Test...
```

Enter the values of yellow candies found in the sample, total number of candies in the sample, and the confidence level.

```
1-PropZInt
  X : 25
  n : 100
  C-Level : 95
  Calculate
```

Arrow down to Calculate, and then press $\boxed{\text{ENTER}}$. The interval is shown along with the sample proportion and sample size.

```
1-PropZInt
  ( .16513, .33487)
  p̂ = .25
  n = 100
```

Changing the confidence level affects the width of the interval. As the confidence level increases, so does the critical value used to construct the interval. Thus, the higher the confidence level, the larger the margin of error and the wider the interval.

EXAMPLE: A random sample of 100 candies has 25 that are yellow. Construct the 90% and 99% confidence intervals for the true proportion of yellow candies.

ANSWER:

$$90\% : 0.25 \pm 1.645\sqrt{\frac{(0.25)(0.75)}{100}} = (0.17878, 0.32122)$$

$$90\% : 0.25 \pm 2.576\sqrt{\frac{(0.25)(0.75)}{100}} = (0.13846, 0.36154).$$

The 95% confidence interval from the previous example is (0.16513, 0.33487).

Note: The interval becomes wider as you increase the confidence level. The widest interval comes from the highest confidence level.

Sometimes, you are asked to find the minimum sample size required to produce a confidence interval with a particular margin of error. Recall, that the margin of error for a one-sample confidence interval for proportion is $ME = z^* \sqrt{\dfrac{\hat{p}(1-\hat{p})}{n}}$. Solving for n, we get $n = \dfrac{(z^*)^2 \hat{p}(1-\hat{p})}{ME^2}$.

EXAMPLE: What is the minimum number of candies needed in a sample to estimate the true proportion of green candies with a 4% margin of error at 95% confidence? Assume that previous information estimated the proportion of green to be 30%.

ANSWER: The margin of error is $ME = z^* \sqrt{\dfrac{\hat{p}(1-\hat{p})}{n}}$. In this case, the maximum allowable margin of error is 0.04. So, $0.04 \geq 1.960\sqrt{\dfrac{0.3(1-0.3)}{n}}$. Solving for n, $n \geq \dfrac{(z^*)^2 \hat{p}(1-\hat{p})}{ME^2} \Rightarrow n \geq \dfrac{(1.960)^2 0.3(0.7)}{0.04^2} \Rightarrow n \geq 504.21$. The calculations lead to a sample size of at least 504.21, so we round up to the next integer; 505 candies are needed.

Note: When you do not have information about the population proportion on which to base your sample size calculations, use $\hat{p} = 0.5$ as the estimate. This will ensure an appropriate sample size.

In the previous example, if we wanted the margin of error to be 2%, we would find that the sample size becomes 2,017, roughly four times that before. To cut the margin of error in half, and therefore the width of the interval, you must quadruple the sample size.

Confidence Intervals for the Difference Between Two Proportions

The formula to compute a confidence interval for the difference of two population proportions

$$p_1 - p_2 \text{ is } (\hat{p}_1 - \hat{p}_2) \pm z^* \sqrt{\frac{\hat{p}_1(1-\hat{p}_1)}{n_1} + \frac{\hat{p}_2(1-\hat{p}_2)}{n_2}}, \text{ where } \hat{p}_1 \text{ and } \hat{p}_2 \text{ are the respective}$$

sample proportions, n_1 and n_2 are the respective sample sizes, and z^* is the z-critical value.

The assumptions/conditions for inference for a difference between two proportions change slightly from the one-sample procedure:

(1) Both samples must be independent SRSs from the populations of interest.

(2) The population sizes are both at least 10 times as large as the respective sample sizes.

(3) The number of successes and failures in both samples are both at least 10. That is, $n_1\hat{p}_1 \geq 10$, $n_1(1-\hat{p}_1) \geq 10$, $n_2\hat{p}_2 \geq 10$, and $n_2(1-\hat{p}_2) \geq 10$.

EXAMPLE: We are interested in finding the difference in the proportions of high school seniors taking AP Statistics between rural and urban areas of a populous state. An SRS of 250 rural high school seniors was taken and 82 are taking AP Statistics. In urban areas, an SRS of 280 seniors produced 56 taking AP Statistics. Compute a 90% confidence interval for the true difference between the proportions of high school seniors taking AP Statistics in rural and urban areas.

ANSWER: Follow the four-step procedure for inference.

Step 1: State the parameter of interest. The parameter of interest is the difference in proportions $p_1 - p_2$, where p_1 = the proportion of all rural seniors taking AP Statistics in the state and p_2 = the true proportion of all urban seniors taking AP Statistics in the state.

Step 2: Name the inference procedure and check assumptions/conditions. The procedure is a 95% confidence interval for $p_1 - p_2$, as defined above.

(1) The samples must be independent SRSs from the populations of interest. OK— the problem states that we have two independent random samples.

(2) The populations are at least 10 times as large as the sample sizes. OK as long as there are 2,500 rural and 2,800 urban seniors in the state. Since the state is "populous," this is reasonable.

(3) The number of successes and failures in each sample are both at least 10. OK— there are 82 successes and 168 failures in the rural sample; there are 56 successes and 224 failures in the urban sample.

Step 3: Calculate the confidence interval. In our sample $\hat{p}_1 = \dfrac{82}{250} = 0.328$, $\hat{p}_2 = \dfrac{56}{180} = 0.2$, and the z-critical value for 90% confidence is 1.645. Substituting into the formula $(\hat{p}_1 - \hat{p}_2) \pm z^* \sqrt{\dfrac{\hat{p}_1(1 - \hat{p}_1)}{n_1} + \dfrac{\hat{p}_2(1 - \hat{p}_2)}{n_2}}$, we get

$$(0.328 - 0.2) \pm 1.645 \sqrt{\frac{0.328(0.672)}{250} + \frac{0.2(0.8)}{280}}$$

$$0.128 \pm 0.0627$$

$$(0.0653, 0.1907).$$

Step 4: Interpret the results in context. We are 90% confident that the difference between the true proportions of high school seniors who take AP Statistics in rural and urban areas is between 7% and 19%. We could alternately say that the proportion of rural seniors taking AP Statistics is between 7% and 19% higher than the proportion of urban seniors.

Calculator Tip:

Confidence intervals for difference in proportions can be calculated with the graphing calculator.

From the home screen, press [STAT], arrow right to TESTS, and then choose B: 2-PropZInt....

```
EDIT  CALC  TESTS
9↑ 2-SampZInt...
0 : 2-SampTInt...
A : 1-PropZInt...
B: 2-PropZInt...
C : χ²-Tes t...
D : χ² GOF-Test...
E↓ 2 -SampFTest...
```

Input the number of rural seniors taking AP Statistics found in the sample, the rural sample size, the number of total urban seniors taking AP Statistics, the urban sample size, and the confidence level.

```
2-PropZInt
 x1 : 82
 n1 : 250
 x2 : 56
 n2 : 280
 C-Level : 90
 Calculate
```

Arrow down to Calculate, and then press [ENTER]. The interval is shown along with the sample proportions and sample sizes.

```
2-PropZInt
 (.0653, .1907)
 p̂1=.328
 p̂2=.2
 n1=250
 n2=280
```

Confidence Intervals for Means

Confidence intervals for population means can be calculated and interpreted as easily as those for proportions. Since we usually do not know the population standard deviation σ, the sample standard deviation s is used to compute standard error. A t-distribution is appropriate to use for the critical value instead of the standard normal when σ is unknown. The value of t^* will depend on degrees of freedom and the confidence level chosen.

The formula to compute a confidence interval for a population mean μ is $\bar{x} \pm t^* \dfrac{s}{\sqrt{n}}$,

where $\bar{x}$ is the sample mean, s is the sample standard deviation, n is the sample size, and t^* is the t-critical value. The number of degrees of freedom is $n-1$. (If the population standard deviation is known, use σ and z^*.)

The assumptions/conditions for inference for a population mean are as follows:

(1) The sample must be an SRS from the population of interest.

(2) The data come from a normally distributed population. This is rarely ever going to be the case. But, we can satisfy this condition if the following criteria are met:

- For small samples (up to size 15), a plot of the data is generally symmetric, unimodal, and has no outliers.

- For sample sizes between 15 and 40, the plot should be unimodal and have no outliers. Moderate skewness can be tolerated.

- For large samples (over size 40), the procedures are useful even in the presence of extreme skewness. Outliers are still a concern.

EXAMPLE: A popular weight loss plan claims that most people will lose an average of 6 pounds per month if they follow the plan exactly as prescribed and exercise 30 minutes each day. A random sample of 45 people who have been on the plan for 1 month produced a mean weight loss of 5.6 pounds and the standard deviation was 1.32 pounds. A dotplot of the data is shown below.

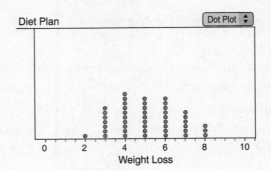

Construct a 95% confidence interval for the true mean weight loss after 1 month on this plan.

◼ ANSWER: Follow the four-step inference procedure.

Step 1: State the parameter of interest. The parameter of interest is μ, the true mean weight loss after 1 month on this diet plan.

Step 2: Name the inference procedure and check assumptions/conditions. The procedure is a 95% confidence interval for μ, as defined above.

(1) The sample must be an SRS from the population of interest. OK—the problem states that we have a random sample.

(2) The data come from a normally distributed population. OK—the dotplot of the data is unimodal and symmetric, with no outliers.

Step 3: Calculate the confidence interval. In our sample $\bar{x} = 5.6$, $s = 1.32$, and the t-critical value for 95% confidence and $n - 1 = 44$ degrees of freedom is 2.021. (Note: If the number of degrees of freedom is not shown in Table B, use the next *lower* value.) Substituting into the formula $\bar{x} \pm t^* \dfrac{s}{\sqrt{n}}$, we get

$$\bar{x} \pm t^* \frac{s}{\sqrt{n}}$$

$$5.6 \pm 2.021 \frac{1.32}{\sqrt{45}}$$

$$5.6 \pm 0.3966$$

$$(5.2034,\ 5.9966).$$

Step 4: Interpret the results in context. We are 95% confident that the true mean weight loss per month for people on this plan is between 5.2 and 6.0 pounds.

Calculator Tip:

Confidence intervals for means can be calculated with the graphing calculator.

From the home screen, press $\boxed{\text{STAT}}$, arrow right to TESTS, and then choose 8:TInterval....

```
EDIT  CALC  TESTS
5↑1-PropZTest...
6 : 2-PropZTest...
7 : ZInterval...
8 : TInterval...
9 : 2-SampZInt...
0 : 2-SampTInt...
A↓ 1-PropZInt...
```

Input the mean and standard deviation of the sample, the sample size, and the confidence level.

```
TInterval
  Inpt : Data  Stats
  x̄ : 5.6
  Sx : 1.32
  n : 45
  C-Level : 95
  Calculate
```

Arrow down to Calculate, and then press $\boxed{\text{ENTER}}$. The interval is shown along with the sample mean, sample standard deviation, and sample size.

```
TInterval
  (5.2034, 5.9966)
  x̄=5.6
  Sx=1.32
  n=45
```

Confidence Intervals for Means with Paired Data

The same formula can be used to estimate mean difference in a paired situation. The matched-pairs design uses a one-sample procedure. Remember that you are using the differences in the two sets of data to find the confidence interval for the mean difference. You do not have to worry about the condition of independence, which we will find is true of two-sample procedures, since the data are paired and not independent of each other.

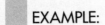 **EXAMPLE:** A baseball bat manufacturer makes two different types of bats, aluminum and ceramic. The company wants to know which type of bat provides better performance for little leaguers as measured by number of balls hit into the outfield. Twenty-three boys who play little league baseball participated in the study. Each player was randomly assigned one type of bat with which he took 20 swings, followed by 20 swings with the other bat. The number of hits into the outfield was recorded for each bat. The number of hits into the outfield for each type of bat is shown in the table below.

Aluminum	4	2	3	3	6	3	5	7	5	9	6	3
Ceramic	5	5	3	6	8	6	6	12	8	15	8	2
Difference (C – A)	1	3	0	3	2	3	1	5	3	6	2	–1
Aluminum	1	2	4	8	8	3	2	2	6	5	4	
Ceramic	3	5	8	7	8	6	5	4	11	9	6	
Difference (C – A)	2	3	4	–1	0	3	3	2	5	4	2	

Find the 90% confidence interval for the mean difference in the number of hits into the outfield between the two types of bats.

 ANSWER: Follow the four-step inference procedure.

Step 1: State the parameter of interest. The parameter of interest is μ_d, the true mean difference in hits into the outfield between ceramic and aluminum bats.

Step 2: Name the inference procedure and check assumptions/conditions. The procedure is a 90% confidence interval for μ_d, as defined above.

(1) The sample must be an SRS from the population of interest. OK—the players may not be a random sample, but the data come from a randomized comparative experiment, so the randomization condition is satisfied.

(2) The data come from a normally distributed population. OK—the histogram of differences is unimodal and symmetric, with no outliers.

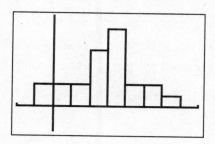

Step 3: Calculate the confidence interval. In our sample $\bar{x}_d = 2.3913$, $s = 1.827$, and the t-critical value for 90% confidence and $n - 1 = 22$ degrees of freedom is 1.717.

Substituting into the formula $\bar{x}_d \pm t^* \dfrac{s_d}{\sqrt{n}}$, we get

$$\bar{x}_d \pm t^* \frac{s_d}{\sqrt{n}}$$

$$2.3913 \pm 1.717 \frac{1.828}{\sqrt{23}}$$

$$2.3913 \pm 0.6544$$

$$(1.7369, 3.0457).$$

Step 4: Interpret the results in context. We are 90% confident that for little league players, the ceramic bat produces, on average, between 1.7 and 3.0 more hits to the outfield (per 20 swings) than the aluminum bat.

Below is an example of a computer printout that you might see. In this case the interval has been calculated by the software and is circled.

Variable	N	Mean	StDev	SE Mean	90.0 % CI
Difference	23	2.391	1.828	0.381	1.737, 3.046

Sometimes you might be given the information with the confidence interval information missing, like that below.

Variable	N	Mean	StDev	SE Mean	
Difference	23	2.391	1.828	0.381	$\dfrac{s}{\sqrt{n}}$

You would be expected to generate an interval from this information. In this case you need the critical value, the sample mean, and the standard deviation of the sample in order to compute a confidence interval. The "SE Mean" on the printout (circled) is the standard error—the standard deviation from the sample divided by the square root of the sample size. The interval would be $2.391 \pm 1.717(0.381)$.

Just as we saw with proportions, we can also determine a minimum sample size needed for a predetermined margin of error. Recall, that the margin of error for a one-sample confidence interval for mean is $ME = z^* \dfrac{\sigma}{\sqrt{n}}$. Solving for n, we get

$$n = \frac{(z^*)^2 \sigma^2}{ME^2}.$$

EXAMPLE: Ball bearings must be manufactured to strict tolerances or they will not function properly. If a quality control inspector randomly samples ball bearings from a production process, what is the minimum sample size required to be 99% confident that the mean diameter is within 0.04 mm? Past data indicate that the standard deviation of bearing diameters is 0.10 mm.

ANSWER: The margin of error is $ME = z^* \dfrac{\sigma}{\sqrt{n}}$. In this case, the maximum allowable margin of error is 0.04 mm. So, So, $0.04 \geq 2.576 \dfrac{0.10}{\sqrt{n}}$. Solving for n,

$$n \geq \frac{(z^*)^2 \sigma^2}{ME^2} \Rightarrow n \geq \frac{(2.576)^2 (0.1)^2}{0.04^2} \Rightarrow n \geq 41.4736.$$

Round up to next integer; we need to sample 42 ball bearings for the required margin of error.

Confidence Intervals for the Difference Between Two Means

A two-sample confidence interval for the difference between population means is used when you have two independent random samples. The formula to compute the interval for $\mu_1 - \mu_2$ is $(\bar{x}_1 - \bar{x}_2) \pm t^* \sqrt{\dfrac{s_1^2}{n_1} + \dfrac{s_2^2}{n_2}}$, where $\bar{x}_1$ and $\bar{x}_2$ are the sample means, s_1 and s_2 are the sample standard deviations, n_1 and n_2 are the sample sizes, and t^* is the t-critical value.

The number of degrees of freedom is computed by a complex formula not presented here. If technology is used, this exact number of degrees of freedom will be provided. If by-hand calculations are performed, a conservative interval can be constructed using the smaller of $n_1 - 1$ and $n_2 - 1$.

If the population standard deviations are known, use σ_1, σ_2, and z^*.

The assumptions/conditions for inference for the difference in population means are as follows:

(1) The samples must be SRSs from the populations of interest.

(2) The data come from normally distributed populations. This is rarely ever going to be the case. But, we can satisfy this condition if the following criteria are met:

- For small samples (up to size 15), plots of the data are generally symmetric, unimodal, and have no outliers.

- For sample sizes between 15 and 40, the plots should be unimodal and have no outliers. Moderate skewness can be tolerated.

- For large samples (over size 40), the procedures are useful even in the presence of extreme skewness. Outliers are still a concern.

EXAMPLE: A cruise line is trying to determine if there is a difference in the mean amount of money passengers spend on board their ships during a 1-week cruise to Alaska and a 1-week cruise to the Caribbean. This includes food, beverages, gift shops, and the spa. Some items are the same on both ships and there are regional souvenirs that are different. Random samples of 35 passengers from each ship are taken. The mean amount of money spent on board the ship in Alaska is $\bar{x}_1 = \$582.16$ with a standard deviation of $s_1 = \$32.65$. The mean amount of money spent on board the ship in the Caribbean is $\bar{x}_2 = \$493.62$ with a standard deviation of $s_2 = \$28.73$. Find the 95% confidence interval for the difference in the mean amounts of money spent on board the ships. Histograms of the data are shown below.

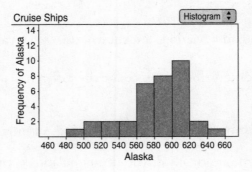
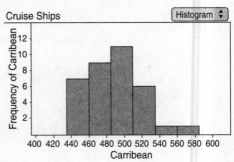

ANSWER: Follow the four-step inference procedure.

Step 1: State the parameter of interest. The parameter of interest is $\mu_1 - \mu_2$, the true difference in the mean amounts spent by passengers on Alaska and Caribbean ships, where μ_1 is the true mean amount spent by Alaskan passengers and μ_2 is the true mean amount spent by Caribbean passengers.

Step 2: Name the inference procedure and check assumptions/conditions. The procedure is a 95% confidence interval for $\mu_1 - \mu_2$, as defined above.

(1) The samples must be SRSs from the populations of interest. OK—the problem states that the samples were random.

(2) The data come from normally distributed populations. OK—the histogram of Caribbean passenger spending is symmetric and unimodal; the histogram of the Alaskan passengers' expenditures is skewed left, but the sample size of 35 is large enough to tolerate it.

Step 3: Calculate the confidence interval. In our samples $\bar{x}_1 = \$582.16$, $s_1 = \$32.65$, $\bar{x}_2 = \$493.62$, $s_2 = \$28.73$, and the *t*-critical value for 95% confidence and 30 degrees of freedom is 2.042. ($df = 34$ is not in the table.)

Substituting into the formula $(\bar{x}_1 - \bar{x}_2) \pm t^* \sqrt{\dfrac{s_1^2}{n_1} + \dfrac{s_2^2}{n_2}}$, we get

$$(\bar{x}_1 - \bar{x}_2) \pm t^* \sqrt{\frac{s_1^2}{n_1} + \frac{s_2^2}{n_2}}$$

$$(582.16 - 493.62) \pm 2.042 \sqrt{\frac{32.65^2}{35} + \frac{28.73^2}{35}}$$

$$88.54 \pm 15.01$$

$$(73.53, 103.55).$$

Step 4: Interpret the results in context. We are 95% confident that the difference in the mean amount of money spent by passengers on board the two ships is between \$73.53 and \$103.55. Another way to interpret this is that passengers on Alaskan cruises spend on average between \$73.53 and \$103.55 more than passengers on Caribbean cruises.

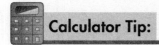

Calculator Tip:

Confidence intervals for the difference in means can be calculated with the graphing calculator.

From the home screen, press $\boxed{\text{STAT}}$, arrow right to TESTS, and then choose 0:2-SampTInt....

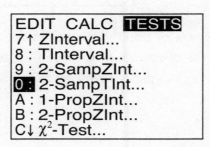

Input the means and standard deviations of the samples, and the sample sizes.

```
2-SampTInt
 Inpt : Data  [Stats]
 x̄1 : 582.16
 Sx1 : 32.65
 n1 : 35
 x̄2 : 493.62
 Sx2 : 28.73
↓n2 : 35
```

Arrow down to input the confidence level. When you reach the **Pooled** line, choose **No**. We only use a pooled t-test if we know the two populations have equal standard deviations, that is, $\sigma_1 = \sigma_2$. This is unlikely to occur.

```
2-SampTInt
↑n1 : 35
 x̄2 : 493.62
 Sx2 : 28.73
 n2 : 35
 C-Level : 95
 Pooled :[No] Yes
 Calculate
```

Arrow down to **Calculate**, and then press [ENTER]. The interval is shown along with the sample means, sample standard deviations, and the exact number of degrees of freedom.

```
2-SampTInt
 (73.866, 103.21)
 df=66.91712053
 x̄1=582.16
 x̄2= 493.62
 Sx1=32.65
↓Sx2=28.73
```

Note: The degrees of freedom used from the table was 30. Calculators and software packages will use the exact degrees of freedom, in this case about 67. The calculator's interval is a bit narrower than the one done by hand.

Confidence Intervals for the Slope of the Regression Line

As we can never truly know the parameters for the true regression line, these being the slope and intercept, we can estimate what these values are. In AP Statistics, we only concern ourselves with the more important of the two: slope. The slope of the regression line is the mean rate of change of the response variable when the explanatory variable increases by one unit. We can estimate the slope of the population model with a confidence interval.

The formula to compute a confidence interval for a population slope β_1 is $b_1 \pm t^* SE_{b1}$, where b_1 is the slope of the sample regression line, SE_{b1} is the standard error of the slope, and t^* is the t-critical value. The number of degrees of freedom is $n-2$.

The assumptions/conditions for inference for the slope of a population model are as follows:

(1) The mean y values for all the fixed x values are related linearly by the equation $\mu_y = \beta_0 + \beta_{yx}$ We check this by verifying that our scatterplot is linear and that a residual plot has no "U"-shaped pattern.

(2) For any fixed value of x, the value of each y is independent. We check this by looking for grouping patterns in the residual plot.

(3) For any fixed value of x, the value of y is normally distributed. We check this by looking at a histogram of the residuals. The plot should be symmetric and uni-modal, although skewness can be tolerated for larger samples.

(4) For all fixed values of x, the standard deviation of y is equal. We check this by verifying that the variability of the residuals is relatively constant across the explanatory variable.

Often, a computer printout will be given to you to calculate the confidence interval for the slope of the true regression model.

EXAMPLE: The heights (in inches) x, and shoe sizes y, of 15 women are shown in the table below.

x	63	62	62	63	68	65	68	69	70	62	60	60	61	63	64
y	7.5	7	7.5	6.5	8	7.5	9	9	10	6	5	5.5	6	7	7.5

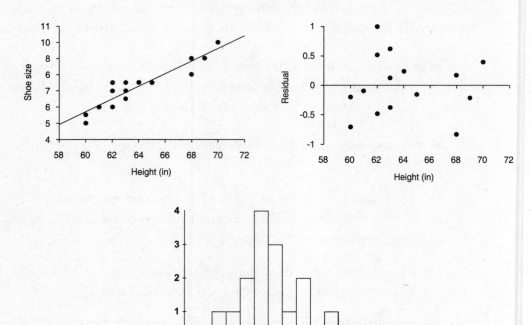

The regression equation is
Shoe Size = - 17.7 + 0.390 Height

Predictor	Coef	StDev	T	P
Constant	-17.693	2.760	-6.41	0.000
Height	0.39000	0.04308	9.05	0.000

S = 0.5276   R-Sq = 86.3%   R-Sq(adj) = 85.3%

slope

SE of slope

Computer output of the regression analysis, a scatterplot, a residual plot, and a histogram of the residuals are provided. Calculate the 95% confidence interval for the slope of the true regression line.

■ ANSWER: Follow the four-step inference procedure.

Step 1: State the parameter of interest. The parameter of interest is β_1, the slope of the population regression model.

Step 2: Name the inference procedure and check assumptions/conditions. The procedure is a 95% confidence interval for β_1, as defined above.

(1) The mean y values for all the fixed x values are related linearly by the equation $\mu_y = \beta_0 + \beta_{1x}$. OK—the scatterplot is linear and the residual plot shows no U-shaped pattern.

(2) For any fixed value of x, the value of each y is independent. OK—there does not appear to be any clustering in the residual plot.

(3) For any fixed value of x, the value of y is normally distributed. OK—the histogram of the residuals is symmetric and unimodal.

(4) For all fixed values of x, the standard deviation of y is equal. OK—the variability of the residuals across the values of x is consistent.

Step 3: Calculate the confidence interval. In our sample $b_1 = 0.39$, $SE_{b1} = 0.043$ and the t-critical value for 95% confidence and $n - 2 = 13$ degrees of freedom is 2.160.

Substituting into the formula $b_1 \pm t^* SE_{b_1}$, we get

$$b_1 \pm t^* SE_{b_1}$$
$$0.39 \pm 2.160(0.04308)$$
$$0.39 \pm 0.093$$
$$(0.297, 0.483).$$

Step 4: Interpret the results in context. We are 95% confident that the slope of the true regression line for height and shoe size is between 0.2969 and 0.4831. That is, for each additional inch of height, a woman's shoe size is between about 0.3 and 0.5 sizes larger.

Calculator Tip:

Confidence intervals for the slope of the population model can be calculated with the graphing calculator. You must, however, have the raw data in lists.

In this case, the women's heights are in list L1 and shoe sizes are in list L2. From the home screen, press ENTER, arrow right to TESTS, and then choose G:LinRegTInt....

```
EDIT  CALC  TESTS
B↑ 2-PropZInt...
C : χ²₂-Test...
D : χ² GOF-Test...
E : 2-SampFTest...
F : LinRegTTest...
G : LinRegTInt...
A : ANOVA (
```

Input the data lists, the confidence level, and where the regression equation is to be stored.

```
LinRegTInt
  Xlist : L₁
  Ylist : L₂
  Freq : 1
  C-Level : 95
  RegEQ : Y1
  Calculate
```

Arrow down to Calculate, and then press ENTER. The interval is shown along with the sample slope, degrees of freedom, standard error about the regression line, and intercept. Scrolling down will give r and r^2.

```
LinRegTInt
  y = a+bx
  ( .29694, .48306)
  b = .39
  df = 13
  s = .5275730597
↓ a = −17.69333333
```

TEST TIP

In order to get the best possible score on the free-response questions in Section II, be sure to provide written explanations for each step of your solution and your final conclusion. Justifying your answer throughout will help the readers get a better grasp of your understanding of statistics, and may help you receive partial credit if you make a computational or other practical error in your work.

Tests of Significance

Significance tests are performed to determine whether the observed value of a sample **statistic** differs significantly from the **hypothesized value** of a population **parameter**. A significance test evaluates evidence with respect to two competing hypotheses: the null hypothesis and the alternative hypothesis. These are both decided before data are collected.

The Two Hypotheses

A **null hypothesis** is written as the claim about the population parameter that is initially thought to be true. It is the hypothesis of no difference, of no change, of no association, or of the status quo. It is a statement of equality. It is always written in the form

H_0: *parameter* = hypothesized value.

The **alternative hypothesis** is tested *against* the null hypothesis, and is based on what the researcher is seeking evidence of. It is a statement of inequality. It can be written looking for the difference or change in one direction from the null hypothesis or both. It is written in one of three forms:

H_a: *parameter* < hypothesized value (one-sided test),

H_a: *parameter* > hypothesized value (one-sided test), or

H_a: *parameter* ≠ hypothesized value (two-sided test).

EXAMPLE: A high school junior running for student body president claims that 80% of the student body favors her in the school election. If her opponent believes this percentage to be lower, write the appropriate null and alternative hypotheses.

ANSWER: The null hypothesis is the statement of no difference and is H_0: $p = 0.80$. The opponent seeks evidence of a lower proportion of votes, so the alternative hypothesis is H_a: $p < 0.80$.

EXAMPLE: In the year 1976, the mean beef consumption of U.S. residents was 89 pounds per person. A health studies researcher wishes to know if this has changed and will do a study. Write the appropriate null and alternative hypotheses to test whether mean beef consumption has changed.

ANSWER: The null hypothesis is the statement of no difference and is H_0: $\mu = 89$ pounds. The researcher seeks evidence of a change in consumption, so the alternative hypothesis is H_a: $\mu \neq 89$ pounds.

Statistical Significance and *P*-Values

When the sample statistic is shown to be far from the hypothesized parameter, the difference is said to be **statistically significant** and we have evidence to **reject** the null hypothesis in favor of the alternative hypothesis. If the statistic is not far, the lack of evidence means we **fail to reject** (*not* accept) the null hypothesis.

We measure the *strength* of the evidence with the **P-value**. The *P*-value tells us how likely it would be to observe a statistic as far as it is, or farther, from the hypothesized value in the direction of the alternative hypothesis, if the null hypothesis were indeed true. The *P*-value is therefore a probability. The smaller the *P*-value, the greater the evidence is *against* the null hypothesis.

EXAMPLE: A high school junior running for student body president claims that 80% of the student body favors her in the school election. Her opponent believes this percentage to be lower and conducts a random sample of likely voters. The opponent tests H_0: $p = 0.80$ versus H_a: $p < 0.80$, where p is the true proportion of voters that favor the first student. The sample results in 72% favoring the first student and the test produces a P-value of 0.02. Interpret the P-value.

ANSWER: The P-value of 0.02 means that if the first candidate really were favored by 80% of the voters, a sample proportion of 72% or less would occur about 2% of the time by chance alone. That is, if H_0 were true, and repeated samples were taken, only 2% of the sample proportions would be 72% or less. This is not very likely and would lead us toward rejecting the null hypothesis.

Sometimes, P-values are compared against a predetermined benchmark, known as the **significance level** or **alpha level**. Three common alpha levels in testing are $\alpha = 0.10$, $\alpha = 0.05$, and $\alpha = 0.01$. If the P-value is less than alpha ($P < \alpha$), we consider the result to be statistically significant and reject the null hypothesis. In the previous example, the results of the survey were statistically significant for $\alpha = 0.10$ and $\alpha = 0.05$, since the P-value is 0.02, but not for $\alpha = 0.01$.

As stated, the P-value is computed by determining the probability of observing a **test statistic** as extreme as or more extreme than we observed if the null hypothesis is true. The test statistic is the number of standard deviations (or standard errors) our sample statistic lies from the population parameter. The P-value's probability is always calculated by finding the area in the tails of the appropriate distribution in the direction(s) of the alternative hypothesis.

EXAMPLE: In 1976, the mean beef consumption of U.S. residents was 89 pounds per person. A health studies research wishes to know if this has decreased and does a study to test H_0: $\mu = 89$ pounds versus H_a: $\mu < 89$ pounds. The test statistic is $t = -2.147$ with 30 degrees of freedom. Compute the P-value for this test.

ANSWER: The test statistic is a t-statistic, so consult Table B: t-distribution critical values. We see that on the $df = 30$ line, $t = 2.147$ corresponds to a tail probability of 0.02. The P-value is 0.02.

Note: If this were a two-sided test, using $H_a: \mu \neq 89$ pounds, we would be interested in the chance of getting a test statistic of -2.147 or less, or $+2.147$ or more. In that case, the P-value would be twice as large, or 0.04.

Decision Errors in Hypothesis Tests

Decision errors can be made when rejecting or failing to reject the null hypothesis. If one sets the significance level to be 0.05, then 5% of the time the null hypothesis is true one will still reject it. Rejecting the null hypothesis when in fact it is true is a **Type I error**. The significance level α is the probability of committing a Type I error. If we fail to reject the null hypothesis in favor of the alternative when in fact the null hypothesis was false, we have committed a **Type II error**. The probability of committing the Type II error is given by the value β. The table below is a summary of decisions that can be made in a hypothesis test.

	What is true about the population parameter	
	H_0 is true	H_a is true
Reject H_0	Type I error probability $= \alpha$	Decision is correct
Fail to reject H_0	Decision is correct	Type II error probability $= \beta$

Decision made based on the sample

It is important to be able to describe Type I and Type II errors in the context of a situation.

EXAMPLE: A manufacturer of computer monitors receives shipments of LCD panels from a supplier overseas. It is not cost effective to inspect each LCD panel for defects, so a sample is taken from each shipment. A significance test is conducted to determine whether the proportion of defective LCD panels is greater than the acceptable limit of 1%. If it is, the shipment will be returned to the supplier. Essentially, this is a test of $H_0: p = 0.01$ vs. $H_a: p > 0.01$, where p is the true proportion of defective panels in the shipment.

(a) Describe the Type I and Type II errors in the context of the situation.

(b) If you were the supplier of the LCD panels, which error would be more serious and why?

(c) If you were the computer monitor manufacturer, which error would be more serious and why?

ANSWER: (a) If a Type I error were committed, we would conclude that there are more than 1% defective panels when there really were not. The shipment would be returned when it was not warranted. If a Type II error were committed, we would conclude that there are no more than 1% of defective panels when there really were. Faulty LCD panels would be accepted from the supplier.

(b) The supplier would think that the Type I error is more serious because they would be receiving LCD panels back that work fine.

(c) The computer monitor manufacturer would think that the Type II error is more serious because they would be receiving panels of poor quality.

Power

The **power** of a test is defined as the probability that you will correctly reject the null hypothesis when it is false. Given β is the probability that you fail to reject the null hypothesis when it is actually false, the power is simply $1 - \beta$. We want the value of the power to be high since that gives us the confidence we made the correct decision concerning the null hypothesis.

You will not be asked to compute power on the AP Statistics Examination, but you must know what factors increase or decrease the power of a test.

(1) Sample size. As the sample size increases, the power of a test increases.

(2) Significance level α. Higher levels of α increase power, as it is more likely that one will reject the null hypothesis.

(3) The value of an alternative parameter. If the alternative hypothesis is true, then the value of the parameter must be something other than what is stated in the null hypothesis. The farther an alternative parameter from the hypothesized value *in the direction of the alternative hypothesis,* the greater the power. This is because a greater difference is easier to detect.

General Procedure for Inference with Hypothesis Tests

To do any inference problem, the same four-step procedure for inference is used. Differences from the confidence interval procedures are shown in *italics.*

Step 1: State the parameter of interest *and a correct pair of hypotheses*. The parameter of interest is from the population under study. It will be a population proportion, a difference of population means, and so on. This should be defined in words and symbols.

The null and alternative hypotheses for the test are written. These may be in the form of words and/or symbols, but if in symbols, those symbols should be defined.

Step 2: Name the inference procedure and check assumptions/conditions. The procedure is named by giving the name of the test and what parameter is being tested.

There are assumptions/conditions that must be checked in order to proceed with various inference procedures. It is not enough to merely state them—you must verify that they apply.

Step 3: Calculate the *test statistic and P-value*. With the exception of chi-square tests (which will be covered later in this chapter), test statistics are always computed as

$$\text{test statistic} = \frac{\text{statistic} - \text{parameter}}{\text{standard deviation of statistic}}.$$

Write the appropriate formula for the type of test statistic being computed, substitute the values into the formula, and compute the test statistic.

Once the test statistic is computed, compute the *P*-value using the direction dictated by the alternative hypothesis and appropriate distribution (z, t, or χ^2).

Step 4: Interpret the results in context. Describe in the context of the situation what the *P*-value means and whether the decision is to reject or fail to reject the null hypothesis. Refer back to step 1, as this is where the parameter of interest and hypotheses were defined.

Hypothesis Tests for Proportions

The formula to compute the test statistic for the population proportion p, with the null hypothesis H_0: $p = p_0$, is $z = \dfrac{\hat{p} - p_0}{\sqrt{\dfrac{p_0(1 - p_0)}{n}}}$, where $\hat{p}$ is the sample proportion, p_0 is the hypothesized parameter, n is the sample size, and z is the z-test statistic. The *P*-value will be computed using the normal (z) distribution.

The assumptions/conditions for a hypothesis test for a population proportion are the same as for confidence intervals with one exception: The number of successes and failures based on the hypothesized parameter are both at least 10. That is, $np_0 \geq 10$ and $n(1 - p_0) \geq 10$.

EXAMPLE: A high school junior running for student body president claims that 80% of the student body will vote for her in the school election. Her opponent believes this to be lower and surveys an SRS of 95 likely voters in student body. The opponent finds that 68 out of 95 students surveyed say they will vote for the junior running for student body president. The school has 3,500 students. Perform a hypothesis test and write a conclusion in the context of the situation.

ANSWER: Follow the four-step inference procedure.

Step 1: State the parameter of interest and a correct pair of hypotheses. Let p be the true proportion of students that will vote for the junior candidate.

H_0: $p = 0.80$

H_a: $p < 0.80$.

Alternately, the parameter of interest and the hypotheses statements can be stated together as shown below.

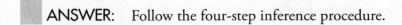

H_0: $p = 0.80$. The true proportion of students that will vote for the junior candidate is 0.80.

$H_a: p < 0.80$. The true proportion of students that will vote for the junior candidate is less than 0.80.

Step 2: Name the inference procedure and check assumptions/conditions. The procedure is a one-sample test for the true proportion of students that will vote for the junior candidate.

(1) The sample must be an SRS from the population of interest. OK—the problem states that we have an SRS.

(2) The population is at least 10 times as large as the sample size. OK—there are more than 950 students in the school. (This may pose a problem if there are *not* 950 likely voters, but we will continue.)

(3) The number of successes and failures based on the hypothesized parameter are both at least 10. OK—since $np_0 = (95)(0.8) = 76$ and $n(1 - p_0) = (95)(0.2) = 19$.

Note: It is required that you show the actual calculations for condition (3). Just restating the formulas will not give credit on the exam.

Step 3: Calculate the test statistic and *P*-value. In our sample $\hat{p} = \dfrac{68}{95}$.

Substituting into the formula $z = \dfrac{\hat{p} - p_0}{\sqrt{\dfrac{p_0(1 - p_0)}{n}}}$, we get

$$z = \frac{\hat{p} - p_0}{\sqrt{\dfrac{p_0(1 - p_0)}{n}}} = \frac{\dfrac{68}{95} - 0.8}{\sqrt{\dfrac{0.8(0.2)}{95}}} \approx -2.05.$$

The *P*-value is the probability of seeing a *z*-test statistic of -2.05 or less, or $P(z \leq -2.05)$. From Table A, this is 0.0202. Thus, *P-value* = 0.0202.

A sketch of the *P*-value is shown below, and is a good idea to show on the exam.

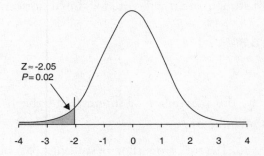

Z ≈ -2.05
P = 0.02

Step 4: Interpret the results in context. The test statistic produced a *P*-value very small (0.02), which means that if 80% of students actually did favor the junior candidate, there is only a 2% chance of getting a test statistic this small or smaller. We have sufficient evidence to reject the null hypothesis in favor of the alternative hypothesis; the true proportion of students who will vote for the junior candidate is less than 0.80.

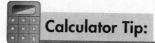 **Calculator Tip:**

Hypothesis tests for proportions can be performed with the graphing calculator.

From the home screen, press $\boxed{\text{STAT}}$, arrow right to TESTS, and then choose 5:1-PropZTest....

```
EDIT  CALC  TESTS
1 : Z-Test...
2 : T-Test...
3 : 2-SampZTest...
4 : 2-SampTTest...
5 : 1-PropZTest...
6 : 2-PropZTest...
7↓ ZInterval...
```

Enter the values of the hypothesized parameter, number of students answering yes (successes), sample size, and highlight the appropriate direction of the alternative hypothesis.

```
1-PropZTest
 p▯ : .8
 x : 68
 n : 95
 prop ≠ p ▯  <p ▯  >p ▯
 Calculate Draw
```

Arrow down to **Calculate**, and then press **ENTER** . The alternate hypothesis, test statistic, *P*-value, sample proportion, and sample size are shown.

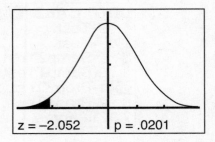

```
1-PropZTest
  prop< .8
  z= −2.051956704
  p= .0200868676
  p̂= .7157894737
  n=95
```

If you had chosen **Draw** instead of **Calculate**, a sketch of the appropriate distribution is shown with the test statistic, *P*-value, and the area that corresponds to the *P*-value is shaded in.

z = −2.052   p = .0201

Hypothesis Tests for the Difference Between Two Proportions

The two-proportion *z*-test is used when you want to know if there is a significant difference between two population proportions. In this case the null hypothesis states that there is no difference between the two population proportions, or $p_1 = p_2$. This can be written alternatively as the difference $p_1 - p_2 = 0$.

The formula to compute the test statistic for the difference of two population proportions $p_1 - p_2$, with the null hypothesis H$_0$: $p_1 - p_2 = 0$, is $z = \dfrac{\hat{p}_1 - \hat{p}_2}{\sqrt{\hat{p}(1 - \hat{p})\left(\dfrac{1}{n_1} + \dfrac{1}{n_2}\right)}}$,

where $\hat{p}_1$ and $\hat{p}_2$ are the respective sample proportions, n_1 and n_2 are the respective sample sizes, $\hat{p}$ is the *pooled* sample proportion, and *z* is the *z*-test statistic. The pooled sample proportion $\hat{p}$ is calculated as $\hat{p} = \dfrac{x_1 + x_2}{n_1 + n_2}$; we "pool" our individual samples and treat them as one big sample. The *P*-value will be computed using the *z*-distribution. The assumptions/conditions for a hypothesis test for a difference in population proportions are the same as for confidence intervals.

EXAMPLE: An airline wants to know if there is a difference in the proportion of business travelers on the 4:00 PM flight between Chicago and New York and the 6:00 PM flight between the same two cities. These flights operate 7 days a week. Random samples of 40 passengers were taken from flight manifests of each flight for the past month. There were 12 business travelers on 4:00 PM flights and 18 on 6:00 PM flights. Is there a significant difference in the proportion of business travelers at these different times?

ANSWER: Follow the four-step inference procedure.

DIDYOUKNOW?

According to the Airports Council International, the world's busiest airport in 2010 was the Atlanta-Hartsfield Airport. Thirteen of the top 30 busiest airports were in the United States, more than any other country.

Step 1: State the parameter of interest and a correct pair of hypotheses. Let p_1 be the true proportion of 4:00 PM business travelers between New York and Chicago. Let p_2 be the true proportion of 6:00 PM business travelers between New York and Chicago. The parameter of interest is $p_1 - p_2$, the true difference in proportion between the two groups.

$$H_0: p_1 - p_2 = 0 \quad \text{or} \quad H_0: p_1 = p_2$$
$$H_a: p_1 - p_2 \neq 0 \quad \quad\quad H_a: p_1 \neq p_2$$

Step 2: Name the inference procedure and check assumptions/conditions. This is a two-proportion *z*-test for $p_1 - p_2$, the true difference in the proportion of business travelers between 4:00 PM and 6:00 PM flights.

(1) The samples must be independent SRSs from the populations of interest. OK—the problem states that we have two independent random samples.

(2) The populations are at least 10 times as large as the sample sizes. OK—it is reasonable to assume that there were at least 400 total passengers on each flight over the course of the month.

(3) The number of successes and failures in each sample are both at least 10. OK—there were 12 successes and 28 failures in the 4:00 PM sample and 18 successes and 22 failures in the 6:00 PM sample.

Step 3: Calculate the test statistic and *P*-value. In our sample

$$\hat{p}_1 = \frac{12}{40} = 0.3, \ \hat{p}_2 = \frac{18}{40} = 0.45, \text{ and } \hat{p} = \frac{12+18}{40+40} = 0.375 .$$

Substituting into the formula, $z = \dfrac{\hat{p}_1 - \hat{p}_2}{\sqrt{\hat{p}(1-\hat{p})\left(\dfrac{1}{n_1} + \dfrac{1}{n_2}\right)}}$, we get

$$z = \frac{\hat{p}_1 - \hat{p}_2}{\sqrt{\hat{p}(1-\hat{p})\left(\dfrac{1}{n_1} + \dfrac{1}{n_2}\right)}} = \frac{0.3 - 0.45}{\sqrt{0.375(0.625)\left(\dfrac{1}{40} + \dfrac{1}{40}\right)}} \approx -1.39.$$

Since the alternative hypothesis is two-sided, the *P*-value is the probability of seeing a *z*-test statistic of -1.39 or less, or $+1.39$ or more. From Table A, $P(z \leq -1.39) = 0.0838$, so the *P*-value $= 2 \times 0.0838 = 0.1676$.

Step 4: Interpret the results in context. The *P*-value found is large—there is about a 17% chance of seeing a test statistic as extreme as or more extreme than the one we observed, if the null hypothesis were true. We fail to reject the null hypothesis. There is not enough evidence to say that there is a difference in the proportion of business travelers between the 4:00 PM and 6:00 PM flights between New York and Chicago.

 Calculator Tip:

Hypothesis tests for differences in proportions can be performed with the graphing calculator.

From the home screen, press STAT , arrow right to TESTS, and then choose 6:2-PropZTest....

```
EDIT  CALC  TESTS
3↑ 2-SampZTest...
4 : 2-SampTTest...
5 : 1-PropZTest...
6 : 2-PropZTest...
7 : ZInterval...
8 : TInterval...
9↓ 2-SampZInt...
```

Input the number of business travelers in the 4:00 PM sample, the 4:00 PM sample size, the number of business travelers in the 6:00 PM sample, the 6:00 PM sample size, and the correct alternative hypothesis that relates p_1 and p_2.

```
2-PropZTest
  x1 : 12
  n1 : 40
  x2 : 18
  n2 : 40
  p1 : ≠ p2 <p2 >p2
Calculate Draw
```

Arrow down to **Calculate**, and then press ENTER. The alternate hypothesis, test statistic, P-value, sample proportions, and pooled proportion are shown. Scrolling down shows the sample sizes.

```
2-PropZTest
  p1 ≠ p2
  z = −1.385640646
  p = .1658567696
  p̂1 = .3
  p̂2 = .45
↓ p̂ = .375
```

Note that the P-value on the calculator screen is very close to what was found by doubling the value from the table. The calculator automatically doubles the table value.

Using the **Draw** feature, we can see that a sketch of the appropriate distribution is shown with the test statistic, P-value, and the area that corresponds to the P-value is shaded in.

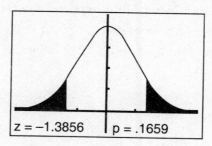

Hypothesis Tests for Means

Hypothesis tests for means have the same issues as confidence intervals. Since we usually do not know the population standard deviation σ, the sample standard deviation s is used to compute standard error.

The formula to compute the test statistic for the population mean μ, with the null hypothesis $H_0: \mu = \mu_0$, is $t = \dfrac{\bar{x} - \mu_0}{\dfrac{s}{\sqrt{n}}}$, where $\bar{x}$ is the sample mean, μ_0 is the hypothesized parameter, s is the sample standard deviation, n is the sample size, and t is the t-test statistic. The number of degrees of freedom is $n - 1$. The P-value will be computed using the t-distribution. (If the population standard deviation is known, use s and the z-distribution.)

The assumptions/conditions for a hypothesis for a population mean are the same as for confidence intervals.

EXAMPLE: At North High School the average number of cars owned by a family is 2.35. South High School is in a more affluent area of the town and students claim they have on average more cars per family. An SRS of 36 students was selected and they were asked how many cars are in each of their families. The number of cars is shown below.

| 1 | 3 | 3 | 4 | 3 | 4 | 5 | 3 | 2 | 3 | 2 | 2 | 3 | 3 | 4 | 4 | 4 | 5 |
| 2 | 1 | 3 | 3 | 2 | 2 | 3 | 3 | 4 | 3 | 4 | 3 | 2 | 2 | 2 | 1 | 1 | 3 |

The mean number of cars from this sample is 2.833 with a standard deviation of 1.056. Is there statistical evidence that there are more cars per family on average at South High School than North High School?

ANSWER: Follow the four-step inference procedure.

Step 1: State the parameter of interest and a correct pair of hypotheses. $H_0: \mu = 2.35$—The mean number of cars per family at South High School is 2.35. $H_0: \mu > 2.35$—The mean number of cars per family at South High School is greater than 2.35.

Step 2: Name the inference procedure and check assumptions/conditions. The procedure is a one-sample t-test for the true mean number of cars per family at South High School.

(1) The sample must be an SRS from the population of interest. OK—the problem states that we have a random sample.

(2) The data come from a normally distributed population. OK—a histogram of the data (shown below) is unimodal and symmetric, with no outliers.

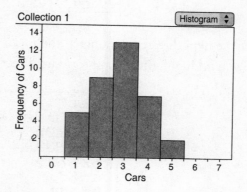

Step 3: Calculate the test statistic and *P*-value. In our sample $\bar{x} = 2.833$ and $s = 1.056$. Substituting into the formula $t = \dfrac{\bar{x} - \mu_0}{\dfrac{s}{\sqrt{n}}}$ we get

$$t = \frac{\bar{x} - \mu_0}{\dfrac{s}{\sqrt{n}}} = \frac{2.833 - 2.35}{\dfrac{1.056}{\sqrt{36}}} \approx 2.74 \text{, with } n - 1 = 35 \text{ degrees of freedom.}$$

Using the t-distribution table with 30 degrees of freedom (since 35 is not in the table, use the next *lower* value), the probability that the test statistic is 2.74 or more is just a little less than 0.005.

Step 4: Interpret the results in context. This extremely small *P*-value is significant for any of the commonly used alpha levels. It means that if the null hypothesis were true—that the mean number of cars owned by families of students at South High School were indeed 2.35—we would see a test statistic *at least* this large about 0.5% of the time, that is about one chance in 200. Thus, we have sufficient evidence to reject the null hypothesis in favor of the alternative one, and we conclude that the true mean number of cars per family at South High School is more than 2.35.

Hypothesis tests for means can be performed with the graphing calculator. From the home screen, press [STAT], arrow right to TESTS, and then choose 2:TTest....

```
EDIT  CALC  TESTS
1↑ Z-Test...
2: T-Test...
3 : 2-SampZTest...
4 : 2-SampTTest...
5 : 1-PropZTest...
6 : 2-PropZTest...
7↓ ZInterval...
```

Input the hypothesized parameter, the mean and standard deviation of the sample, the sample size, and highlight the appropriate direction of the alternative hypothesis.

```
T-Test
 Inpt : Data  Stats
 μ₀ : 2.35
 x̄ : 2.833
 Sx : 1.056
 n : 36
 μ : ≠μ₀   <μ₀   >μ₀
 Calculate Draw
```

Arrow down to **Calculate**, and then press [ENTER]. The alternative hypothesis, t-test statistic, P-value, sample mean, sample standard deviation, and sample size are shown.

```
T-Test
 μ>2.35
 t =2.744318182
 p= .0047507285
 x̄=2.833
 Sx=1.056
 n=36
```

Using the **Draw** feature, we can see that a sketch of the appropriate distribution is shown with the test statistic and P-value, and the area that corresponds to the P-value is shaded in (but is so small that it cannot be distinguished from the curve and axis).

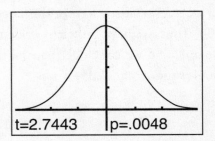

t=2.7443 |p=.0048

You may also be expected to interpret hypothesis test results from a computer printout. The following is a printout from statistics software for the previous example.

T-Test of the Mean

Test of mu = 2.350 vs mu > 2.350

Variable	N	Mean	StDev	SE Mean	T	P-Value
Cars	36	2.833	1.056	0.176	2.75	0.0047

$$\frac{s}{\sqrt{n}} \qquad \text{test statistic}$$

The printout shows the null and alternative hypotheses, the sample information, the standard error of the sample mean $\dfrac{s}{\sqrt{n}}$, the t-test statistic, and the test's P-value.

Hypothesis Tests for Means with Paired Data

Another use of the one-sample t-procedure is looking at paired differences. Often called a **matched-pairs t-test**, it tests to see if there is a significant mean difference between two groups that are paired in some way. The assumptions/conditions for this test and formula for the test statistic are the same as the one-sample t-test; however, the null hypothesis is that there is no mean difference, μ_d.

EXAMPLE: A baseball bat manufacturer makes two different types of bats, aluminum and ceramic. The company wants to know which type of bat provides better performance for little leaguers as measured by the number of balls hit into the outfield. Twenty-three boys who play little league baseball participated in the study. Each player was randomly assigned one type of bat with which he took 20 swings, followed by 20 swings with the other bat. The number of hits into the outfield was recorded for each bat. (The raw data are provided in a table in the previous section on confidence intervals.) Is there statistical evidence that ceramic bats produce more hits into the outfield than aluminum bats?

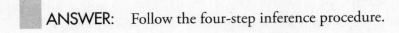ANSWER: Follow the four-step inference procedure.

Step 1: State the parameter of interest and a correct pair of hypotheses.

$H_0: \mu_d = 0$—The mean difference (ceramic − aluminum) in the number of hits into the outfield is 0.

$H_0: \mu_d > 0$—The mean difference (ceramic − aluminum) in the number of hits into the outfield is greater than 0. (We believe they will get more hits, on average, with the ceramic bat)

Step 2: Name the inference procedure and check assumptions/conditions. A one-sample paired t-test for the true mean difference in hits into the outfield between ceramic and aluminum bats.

(1) The sample must be an SRS from the population of interest. OK—the players may not be a random sample, but the data come from a randomized comparative experiment, so the randomization condition is satisfied.

(2) The data come from a normally distributed population. OK—the histogram of differences is unimodal and symmetric, with no outliers.

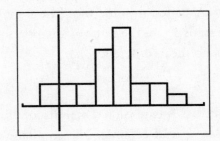

Step 3: Calculate the test statistic and *P*-value. In our sample $\bar{x}_d = 2.3913$ and $s_d = 1.827$.

Substituting into the formula $t = \dfrac{\bar{x} - \mu_0}{\frac{s}{\sqrt{n}}}$, we get $t = \dfrac{\bar{x}_d - \mu_0}{\frac{s_d}{\sqrt{n}}} = \dfrac{2.3913 - 0}{\frac{1.8275}{\sqrt{25}}} \approx 6.28$, with

$n - 1 = 24$ degrees of freedom.

Using the t-distribution table with 24 degrees of freedom, the probability that the test statistic is 6.28 or more is smaller than 0.0005, the last column in the table. We know that $P < 0.0005$.

A calculator output for the test reveals that the *P*-value is 1.37×10^{26}, or 0.00000137. That is about one chance in 730,000!

```
T-Test
 μ>0
 t=6.249145612
 p=1.3679192E–6
 x̄=2.3813
 Sx=1.8275
 n=23
```

Step 4: Interpret the results in context. The test statistic produced an extremely small P-value. There is overwhelming evidence to reject the null hypothesis, and it is concluded that the ceramic bats will produce a higher average number of hits into the outfield than the aluminum bats.

Hypothesis Tests for the Difference Between Two Means

We use a two-sample t-test when we are interested in the difference in the means of two independent populations. In this case the null hypothesis states that there is no difference between the two population means, or $\mu_1 = \mu_2$. This can be written alternatively as the difference $\mu_1 - \mu_2 = 0$.

The formula to compute the test statistic for the difference of the two population means $\mu_1 - \mu_2$, with the null hypothesis $H_0: \mu_1 - \mu_2 = 0$, is

$$t = \frac{\bar{x}_1 - \bar{x}_2}{\sqrt{\dfrac{s_1^2}{n_1} + \dfrac{s_2^2}{n_2}}},$$

where $\bar{x}_1$ and $\bar{x}_2$ are the sample means, s_1 and s_2 are the sample standard deviations, and n_1 and n_2 are the sample sizes. The P-value will be computed using the t-distribution.

As with confidence intervals, the number of degrees of freedom is computed by a complex formula not presented here. If technology is used, the exact number of degrees of freedom will be obtained. If by-hand calculations are performed, use the smaller of $n_1 - 1$ and $n_2 - 1$.

If the population standard deviations are known, use σ_1, σ_2, and the z-distribution.

The assumptions/conditions for a hypothesis test for a difference in population means are the same as for confidence intervals.

EXAMPLE: A cruise line is trying to determine if there is a difference in the mean amount of money passengers spend on board their ships during a 1-week cruise to Alaska and a 1-week cruise to Mexico. A random sample of 35 passengers is taken from the Alaskan cruise; a random sample of 30 passengers is taken from the Mexican cruise. The mean amount of money spent on board the ship in Alaska is $\bar{x}_1 = \$582.16$ with a standard deviation of $s_1 = \$32.65$. The mean amount of money spent on board the ship in Mexico is $\bar{x}_2 = \$563.22$ with a standard deviation of $s_2 = \$35.23$. Is there a significant difference between the mean amounts spent by passengers on the two cruise ships? Histograms of the data are shown below.

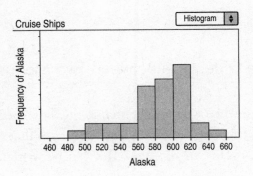 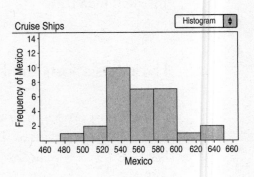

ANSWER: Follow the four-step inference procedure.

Step 1: State the parameter of interest and a correct pair of hypotheses.

$H_0: \mu_1 = \mu_2$ The mean amount of money spent by passengers on board the cruise to Alaska (μ_1) is the same as the mean amount of money spent by passengers on board the cruise to the Mexico (μ_2).

$H_0: \mu_1 \neq \mu_2$ The mean amount of money spent by passengers on board the cruise to Alaska is *not* the same as the mean amount of money spent by passengers on board the cruise to the Mexico.

Step 2: Name the inference procedure and check assumptions/conditions. A two-sample *t*-test of means for the true difference in the mean amount of money spent by passengers on board the two different cruises.

(1) The samples must be SRSs from the populations of interest. OK—the problem states that the samples were random.

(2) The data come from a normally distributed population. OK—the histogram of Mexican passenger spending is symmetric and unimodal; the histogram of the Alaskan passengers' expenditures is skewed left, but the sample size of 35 is large enough to tolerate it.

Step 3: Calculate the test statistic and *P*-value. In our sample $\bar{x}_1 = \$582.16$, $s_1 = \$32.65$, $\bar{x}_2 = \$563.22$, and $s_2 = \$35.23$. Substituting into the formula

$$t = \frac{\bar{x}_1 - \bar{x}_2}{\sqrt{\dfrac{s_1^2}{n_1} + \dfrac{s_2^2}{n_2}}} \text{ we get } t = \frac{\bar{x}_1 - \bar{x}_2}{\dfrac{s_1^2}{n_1} + \dfrac{s_2^2}{n_2}}$$

$$= \frac{582.16 - 563.22}{\sqrt{\dfrac{32.65^2}{35} + \dfrac{35.23^2}{3}}} \approx 2.23 \text{, with } n_2 - 1 = 29 \text{ degrees of freedom.}$$

Using the *t*-distribution table with 29 degrees of freedom, the probability that the test statistic is 2.23 or more is between 0.01 and 0.02. Since this is a two-sided test, the *P*-value is the double of what the table gives, so $0.02 < P < 0.04$.

Step 4: Interpret the results in context. The *P*-value is between 0.02 and 0.04, indicating that if the null hypothesis were really true, we would see a test statistic this or more extreme between 2% and 4% of the time. This is fairly small and gives evidence to reject the null hypothesis.

There is evidence that the mean amount of money spent by passengers on board each cruise ship is different.

Calculator Tip:

Hypothesis tests for the difference in means can be performed with the graphing calculator.

From the home screen, press $\boxed{\text{STAT}}$, arrow right to TESTS, and then choose 4:2-SampTTest....

```
EDIT  CALC  TESTS
1 : Z-Test...
2 : T-Test...
3 : 2-SampZTest...
4  2-SampTTest...
5 : 1-PropZTest...
6 : 2-PropZTest...
7↓ ZInterval...
```

Input the means and standard deviations of the samples, and the sample sizes.

```
2-SampTTest
 Inpt : Data  [Stats]
 x̄1 : 582.16
 Sx1 : 32.65
 n1 : 35
 x̄2 : 563.22
 Sx2 : 35.23
↓n2 : 30
```

Arrow down to select the appropriate alternative hypothesis. When you reach the Pooled line, choose **No**. We only use a pooled *t*-test if we know the two populations have equal standard deviations, that is, $\sigma_1 = \sigma_2$.

```
2-SampTTest
↑n1 : 35
 x̄2 : 563.22
 Sx2 : 35 : 23
 n2 : 30
 μ1 : [≠μ2] <μ2  >μ2
 Pooled : [No] Yes
 Calculate Draw
```

Arrow down to **Calculate**, and then press $\boxed{\text{ENTER}}$. The alternative hypothesis, test statistic, *P*-value, exact number of degrees of freedom, and sample means are shown. Scrolling down gives the sample standard deviations and sample sizes.

```
2-SampTTest
 μ1 ≠ μ2
 t = 2.234747215
 p = .0291836162
 df = 59 78123227
 x̄1 = 582.16
↓x̄2 = 563.22
```

Note: The degrees of freedom used from the table was 29. Calculators and software packages will use the exact degrees of freedom, in this case about 60. The corresponding *P*-value is about 0.03.

Using the **Draw** feature, we can see that a sketch of the appropriate distribution is shown with the test statistic and *P*-value, and the area that corresponds to the *P*-value is shaded.

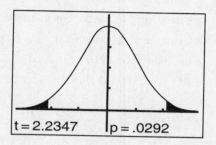

Duality of Confidence Intervals and Hypothesis Tests

Hypotheses can also be evaluated with confidence intervals, provided that the alternative hypothesis is two-sided. If the hypothesized parameter is within a *C*% confidence interval, then there is not statistically significant evidence of a difference at the $\alpha = 1 - C$ significance level. We would fail to reject the null hypothesis in this case. The opposite is true if the hypothesized parameter is not in the interval.

EXAMPLE: An airline wants to know if there is a difference in the proportion of business travelers on the 4:00 PM flight between Chicago and New York and the 6:00 PM flight between the same two cities. The null and alternative hypotheses for a two-sample test of proportions are $H_0: p_1 - p_2 = 0$ and $H_a: p_1 - p_2 \neq 0$, where p_1 is the true proportion of 4:00 PM business travelers between New York and Chicago and p_2 is the true proportion of 6:00 PM business travelers between New York and Chicago. The 95% confidence interval for the true difference in proportions, $p_1 - p_2$, is $(-0.360, 0.060)$. Is there evidence of a significant difference between the two proportions of business travelers?

ANSWER: No, there is insignificant evidence of a difference in the proportion of business travelers on the two flights at the $\alpha = 0.05$ level of significance. The interval contains zero, so it is plausible that there is no difference in the proportion of business flyers. Note that this is consistent with the results of the hypothesis test earlier in the section.

EXAMPLE: Consider the cruise line example from a few pages back. The question asked if there was a significant difference between the mean amount spent by passengers on Alaskan and Mexican cruises. We tested $H_0: \mu_1 = \mu_2$ vs. $H_a: \mu_1 \neq \mu_2$, where μ_1 and μ_2 are the mean amounts of money spent by passengers on board Alaskan and Mexican cruises, respectively. The P-value for the two-sample t-test was 0.03. For which of the common confidence levels, 90%, 95%, and 99%, would the confidence interval for $\mu_1 - \mu_2$ contain zero?

DIDYOUKNOW?

In 2010, Americans took some 1.9 billion trips. Of these, nearly 450 million were for business purposes.

ANSWER: The test is significant at the common significance levels $\alpha = 0.05$ and $\alpha = 0.10$. These correspond to the 95% and 90% confidence intervals, respectively. Therefore, zero would not be in those intervals, and would not be a plausible value for the difference. The test is not significant at the $\alpha = 0.01$ level, so the 99% confidence interval would contain zero. The intervals are provided below as a point of reference.

90% CI: (4.78, 33.10)

95% CI: (1.99, 35.89)

99% CI: (−3.61, 41.49)

One-sided alternatives can also be addressed with confidence intervals, but one-sided intervals are not part of the AP Statistics curriculum.

Inference for Categorical/Count Data

There are several procedures to compare the distribution of categorical data to a hypothesized distribution. Two have already been discussed in this book: the one-sample and two-sample proportion z-tests. These tests work for categorical data of a single variable that is binomial—there are only two outcomes—for inference on one and two populations, respectively.

When a categorical variable has multiple categories, when there are two categorical variables under consideration, or when there are multiple populations under study, a new type of test is required: the chi-square test. The chi-square test is used to compare count data from categorical variables to a hypothesized distribution. There are three forms of chi-square tests, each used in a particular situation.

Chi-square test of goodness of fit: One categorical variable with multiple categories from one population. The test compares the distribution of sample counts with the hypothesized distribution of the population.

Chi-square test of homogeneity: One categorical variable with multiple categories from two or more populations. The test compares the distribution of sample counts with the hypothesized distribution of the population assuming the populations have identical distributions.

Chi-square test of independence: Two categorical variables with multiple categories from one population. The test compares the distribution of sample counts with the hypothesized distribution of the population assuming the two variables are independent.

Chi-Square Test of Goodness of Fit

The chi-square test of goodness of fit is used to compare the distribution of sample counts of one categorical variable from one population with a hypothesized distribution of the population. For instance, a certain brand of bite-sized candies comes in four colors: red, green, blue, and yellow. The manufacturer claims that all colors are made in equal proportion. One particular bag of 60 candies contains 8 red, 16 green, 15 blue, and 21 yellow—certainly not 25% of each color. There is certainly random variability in the process of mixing and packaging the candies, but are these results so far off from 15 of each that the manufacturer's claim can be called into question?

The previous question calls for a chi-square test of goodness of fit. There is one variable (color) with multiple categories (red, green, blue, yellow) and a single population under consideration (all candies of this type by the manufacturer). One will compare the counts of each category observed in the sample to what one would expect to see.

To perform a chi-square test of goodness of fit, the same four-step procedure for inference is used.

Step 1: State the parameter of interest and a correct pair of hypotheses. The null hypothesis is that all proportions of the categories of the variable are not different from what one would expect—the status quo.

$H_0: p_{\text{category 1}}$ = hypothesized proportion of category 1

$p_{\text{category 2}}$ = hypothesized proportion of category 2

$p_{\text{category 3}}$ = hypothesized proportion of category 3

etc.

The alternative hypothesis is that *any one of* the proportions of the categories of the variable is different from what one would expect.

H_a: at least one of the statements in H_0 is not true.

There is no "parameter" of interest. One wants to know about the distribution of proportions over the categories of the variable.

Step 2: Name the inference procedure and check assumptions/conditions. The procedure is the chi-square test of goodness of fit.

There are three assumptions/conditions that must be checked in order to proceed with the chi-square test for goodness of fit.

(1) The data come from a simple random sample.

(2) The size of the sample is no more than 10% of the population size.

(3) All expected cell counts are at least 5.

The expected cell counts for each category of the variable are equal to the product of the sample size and the hypothesized proportion. That is, the expected count of category one $= np_1$, the expected count of category two $= np_2$, and so on.

Step 3: Calculate the test statistic and *P*-value. The chi-square test statistic is computed as

$$\chi^2 = \sum_{\substack{\text{all} \\ \text{cells}}} \frac{(\text{observed} - \text{expected})^2}{\text{expected}}.$$

The number of degrees of freedom for the chi-square test for goodness of fit is equal to one less than the number of categories of the variable.

The *P*-value for the test would be the probability of obtaining a χ^2-statistic at least as large as that computed above if the null hypothesis were true. It is determined by consulting Table C, the table of χ^2 critical values, or by using technology.

Step 4: Interpret the results in context. Make a conclusion as to whether to reject, or fail to reject, the null hypothesis, based on the meaning of the *P*-value. Describe in the context of the situation what the conclusion means.

EXAMPLE: A certain brand of bite-sized candies comes in four colors: red, green, blue, and yellow. The manufacturer claims that all colors are made in equal proportion. One particular bag of 60 candies contains 8 red, 16 green, 15 blue, and 21 yellow. Is there evidence that the proportions of the candy colors differ from the manufacturer's claim?

ANSWER:

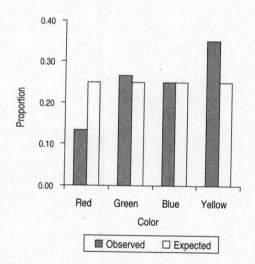

The graph suggests that there may be a difference between the observed proportions of each color and the manufacturer's claimed proportions. Because this difference may be due to sampling variability, an inference procedure is performed to assess the chance of this.

Step 1: State the parameter of interest and a correct pair of hypotheses.

$H_0: p_{red}$ = 0.25

p_{green} = 0.25

p_{blue} = 0.25

p_{yellow} = 0.25

H_a: at least one of the statements in H_0 is not true.

There is no parameter of interest in a chi-square test.

Step 2: Name the inference procedure and check assumptions/conditions. Chi-square test of goodness of fit to determine if the distribution of the proportions of the candy colors matches the manufacturer's claim of equal proportions.

(1) The data come from a simple random sample: OK—the bag of candies is likely a random sample of candies produced.

(2) The size of the sample is no more than 10% of the population size: OK—there are 60 candies in the sample. It is likely that the manufacturer has produced more than 600 candies.

(3) All expected cell counts are at least 5. OK—expected cell counts are given below and all are at least 5.

Actual Counts

Color				
Red	Green	Blue	Yellow	Total
8	16	15	21	60

Expected Counts

Color				
Red	Green	Blue	Yellow	Total
(60)(0.25) = 15	(60)(0.25) = 15	(60)(0.25) = 15	(60)(0.25) = 15	60

Step 3: Calculate the test statistic and _P_-value.

$$\chi^2 = \sum_{\substack{all \\ cells}} \frac{(\text{observed} - \text{expected})^2}{\text{expected}}$$

$$= \frac{(8-15)^2}{15} + \frac{(16-15)^2}{15} + \frac{(15-15)^2}{15} + \frac{(21-15)^2}{15}$$

$$\approx 3.266\ldots + 0.066\ldots + 0 + 2.4$$

$$\approx 5.733.$$

df = number of categories $- 1 = 4 - 1 = 3$.

P-value from Table C: the χ^2-statistic of 5.733 lies between 5.32 and 6.25 in the 3 *df* row, so $0.10 < p < 0.15$.

Step 4: Interpret the results in context. The *P*-value is between 0.10 and 0.15. If the colors were distributed as the manufacturer claims, a chi-square statistic of 5.733 or larger would occur somewhere between 10% and 15% of the time—between 1 chance in 7 and 1 chance in 10. This is fairly likely to occur by chance and leads one to fail to reject the null hypothesis. There is insufficient evidence that the distribution of candy colors differs from that of the manufacturer's claim.

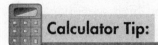

Calculator Tip:

As shown in Chapter 5, the *X*2cdf(function can compute the *P*-value.

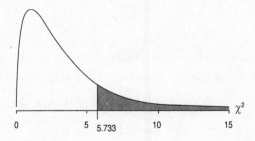

$$X^2cdf(5.733, 1000, 3)$$
$$.1253481055$$

A corresponding sketch of the χ^2-distribution is shown below.

Note: Even though the null hypothesis is defined in terms of the *proportion* of each category, the chi-square test statistic is computed using *counts* of each category.

Chi-Square Test of Homogeneity

The chi-square test of homogeneity is used to compare the distributions of sample counts of one categorical variable from several populations to see if the distributions of the populations are identical. For instance, a certain brand of bite-sized candies comes in three varieties: smooth, crunchy, and chewy. The manufacturer is interested if preferences for the types of candies differ between three school-age groups: elementary, middle, and high school. Random samples of students at three local schools, one of each age group, are taken and the distribution of candy preferences is compared.

The previous situation calls for a chi-square test of homogeneity. There is one variable (preferred variety) with multiple categories (smooth, crunchy, and chewy) and multiple populations under consideration (elementary-, middle-, and high school-aged children). One will compare the counts of each category observed among the different samples to see if they are identically distributed.

To perform a chi-square test of homogeneity, the same four-step inference procedure is used.

Step 1: State the parameter of interest and a correct pair of hypotheses. The null hypothesis is that all proportions of the categories of the variable are the same for all populations (they are homogeneous).

$$H_0: p_{\text{category 1 of population 1}} = p_{\text{category 1 of population 2}} = p_{\text{category 1 of population 3}} = \cdots$$

$$p_{\text{category 2 of population 1}} = p_{\text{category 2 of population 2}} = p_{\text{category 2 of population 3}} = \cdots$$

$$p_{\text{category 3 of population 1}} = p_{\text{category 3 of population 2}} = p_{\text{category 3 of population 3}} = \cdots$$

etc.

The alternative hypothesis is that *any one* of the proportions of the categories of the variable is not the same for all populations under consideration.

H_a: at least one of the statements in H_0 is not true.

There is no parameter of interest. One wants to know about the distributions of proportions over the categories of the variable for each population.

Step 2: Name the inference procedure and check assumptions/conditions. The procedure is the chi-square test of homogeneity.

The assumptions/conditions for the chi-square test for homogeneity are the same as the test for goodness of fit.

The expected cell counts for each cell are equal to $\dfrac{(\text{row toal})(\text{column total})}{\text{grand total}}$.

This is equivalent to the count that each category of the response variable would have if it were proportionally distributed among the samples.

Step 3: Calculate the test statistic and *P*-value. The chi-square test statistic is computed as

$$\chi^2 = \sum_{\substack{\text{all} \\ \text{cells}}} \frac{(\text{observed} - \text{expected})^2}{\text{expected}}.$$

The number of degrees of freedom for the chi-square test for homogeneity is equal to (number of rows $-$ 1)(number of columns $-$ 1).

The *P*-value for the test would be the probability of obtaining a χ^2-statistic at least as large as that computed above, if the null hypothesis were true. It is determined by consulting Table C, the table of χ^2 critical values, or by using technology.

Step 4: Interpret the results in context. Make a conclusion as to whether to reject, or fail to reject, the null hypothesis, based on the meaning of the *P*-value. Describe in the context of the situation what the conclusion means.

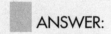 **EXAMPLE:** A certain brand of bite-sized candies comes in three varieties: creamy, crispy, and chewy. The manufacturer is interested if preferences for the types of candies differ between three school-age groups: elementary, middle, and high school. Random samples of students at three local schools, one of each age group, are taken and the sample data compiled in the table below.

		Variety		
		Creamy	Crispy	Chewy
Population	Elementary	33	14	19
	Middle	21	16	17
	High	16	12	32

Is there evidence that the proportion of students that prefer each variety of candy differs among the various age groups?

ANSWER:

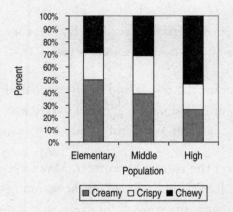

The graph suggests that there may be a difference in the proportions of each variety among the three populations. There seems to be an increase in the popularity of chewy candies and a corresponding decrease in the popularity of creamy candies as students get older. Because the observed difference may be due to sampling variability, an inference procedure is performed to assess the chance of this.

Step 1: State the parameter of interest and a correct pair of hypotheses.

H_0: $p_{\text{creamy in elementary}} = p_{\text{creamy in middle}} = p_{\text{creamy in high}}$

$p_{\text{crispy in elementary}} = p_{\text{crispy in middle}} = p_{\text{crispy in high}}$

$p_{\text{chewy in elementary}} = p_{\text{chewy in middle}} = p_{\text{chewy in high}}$

H_a: at least one of the statements in H_0 is not true.

Step 2: Name the inference procedure and check assumptions/conditions. Chi-square test of homogeneity to determine if the distribution of proportions of preferred candy varieties is the same for all levels of students.

(1) The data come from independent random samples: OK—each group of students was randomly selected from their respective schools.

(2) The size of the samples is no more than 10% of the population sizes: OK so long as there are at least 660 elementary, 540 middle, and 600 high school students in the respective populations.

(3) All expected cell counts are at least 5. OK—expected cell counts are given below and all are at least 5.

Actual Counts

		Variety			
		Creamy	Crispy	Chewy	Total
Population	Elementary	33	14	19	66
	Middle	21	16	17	54
	High	16	12	32	60
	Total	70	42	68	180

Expected Counts

		Variety			
		Creamy	Crispy	Chewy	Total
Population	Elementary	$\dfrac{(66)(70)}{180}$ ≈ 25.667	$\dfrac{(66)(42)}{180}$ $= 15.4$	$\dfrac{(66)(68)}{180}$ ≈ 24.933	66
	Middle	$\dfrac{(54)(70)}{180}$ $= 21$	$\dfrac{(54)(42)}{180}$ $= 12.6$	$\dfrac{(54)(68)}{180}$ $= 20.4$	54
	High	$\dfrac{(60)(70)}{180}$ ≈ 23.333	$\dfrac{(60)(42)}{180}$ $= 14$	$\dfrac{(60)(68)}{180}$ ≈ 22.667	60
	Total	70	42	68	180

$$\text{Expected cell count} = \frac{(\text{row total})(\text{column total})}{\text{grand total}}.$$

Note: The row and column totals for the actual and expected cell counts are equal.

Step 3: Calculate the test statistic and P-value.

$$\chi^2 = \sum_{\substack{\text{all} \\ \text{cells}}} \frac{(\text{observed} - \text{expected})^2}{\text{expected}}$$

$$= \frac{(33 - 25.667)^2}{25.667} + \frac{(14 - 15.4)^2}{15.4} + \frac{(19 - 24.933)^2}{24.933}$$

$$+ \frac{(21 - 21)^2}{21} + \frac{(16 - 12.6)^2}{12.6} + \frac{(17 - 20.4)^2}{20.4}$$

$$+ \frac{(16 - 23.333)^2}{23.333} + \frac{(12 - 14)^2}{14} + \frac{(32 - 22.667)^2}{22.667}$$

$$\approx 2.095 + 0.127 + 1.412$$
$$+ 0 + 0.917 + 0.567$$
$$+ 2.305 + 0.286 + 3.843$$
$$\approx 11.552.$$

$df = (\text{number of rows} - 1)(\text{number of columns} - 1) = (3 - 1)(3 - 1) = 4.$

P-value from Table C: the χ^2-statistic of 11.552 lies between 11.14 and 11.67 in the 4 df row, so $0.02 < p < 0.025$.

Step 4: Interpret the results in context. The P-value is between 0.02 and 0.025. If the preferences for the different varieties of candies were distributed in the same proportions among all three age groups of students, a chi-square statistic of 11.552 or larger would occur somewhere between 2% and 2.5% of the time—between 1 chance in 40 and 1 chance in 50. This is very unlikely to occur by chance and leads one to reject the null hypothesis. There is strong evidence that the preferences for the different varieties of candies are different among the three age groups of students.

The differences in expected and observed counts of Elementary/Creamy, High/Creamy, and High/Chewy appear to be contributing most to the differences between the populations.

TEST TIP

Keep an eye on the clock as you work on your responses. Setting your watch to 12:00 at the beginning of each section can help you quickly see how much time you have used without having to remember your actual start time.

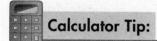 **Calculator Tip:**

The chi-square test of homogeneity can be computed with a graphing calculator. Consider the previous example.

First, the two-way table must be entered into a matrix. From the home screen, press 2nd x^{-1}, which brings up the **MATRIX** screen.

```
NAMES MATH EDIT
1 : [A]
2 : [B]
3 : [C]
4 : [D]
5 : [E]
6 : [F]
7↓[G]
```

Scroll right to the **EDIT** column, and then press $\boxed{\text{ENTER}}$ to select matrix [A].

```
NAMES MATH EDIT
1 : [A]
2 : [B]
3 : [C]
4 : [D]
5 : [E]
6 : [F]
7↓[G]
```

At this point, the dimensions of the matrix are entered as rows □ columns. In this example, there are three rows and three columns; key in $\boxed{3}$ $\boxed{\text{ENTER}}$ $\boxed{3}$ $\boxed{\text{ENTER}}$.

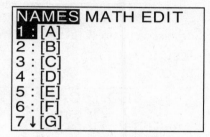

```
MATRIX[A]      3 x 3
[0      0      0      ]
[0      0      0      ]
[0      0      0      ]
```

Enter the values from the two-way table going across each row, column by column, pressing ENTER after each entry.

```
MATRIX[A] 3 x3
[33     14      19      ]
[21     16      17      ]
[16     12      32      ]


3, 3 = 32
```

Return to the home screen, press STAT, scroll right arrow to the TEST menu, and then choose C: χ^2-Test....

```
EDIT CALC TESTS
7↑ ZInterval...
8 : TInterval...
9 : 2-SampZInt...
0 : 2-SampTInt...
A : 1-PropZInt...
B : 2-PropZInt...
C: χ²-Test...
```

This is the chi-square test command for two-way tables. The matrix containing the observed counts is named on the first line.

```
χ²-Test
  Observed : [A]
  Expected : [B]
  Calculate Draw
```

If the correct matrix is not named, key in 2nd x^{-1}, which brings up the MATRIX screen, and select the correct matrix from the NAMES menu. The expected count matrix will be calculated when the test is executed. From the NAMES menu in the MATRIX screen, select which matrix will hold the expected counts.

Scroll down to Calculate and press ENTER .

χ^2 -Test
χ^2 = 11.55219421
p = .0210112805
df = 4

The chi-square test statistic is about 11.552 and the corresponding *P*-value, for a χ^2-curve with 4 degrees of freedom, is approximately 0.021.

To access the expected counts, go to the **MATRIX** screen and select the matrix in which the expected cell counts were stored.

NAMES MATH EDIT
1 : [A]   3x3
2 : [B]   3x3
3 : [C]
4 : [D]
5 : [E]
6 : [F]
7↓[G]

Press ENTER twice.

[B]
[[25.66666667 1...
[21   1...
[23.33333333 1...
■

The expected cell counts are displayed on the screen. You may have to scroll right if the numbers run off the screen.

Repeating the above steps, but choosing Draw from the *X2*-Test function instead of **Calculate** displays the χ^2-curve with the appropriate shaded area, the test statistic, and the *P-value*.

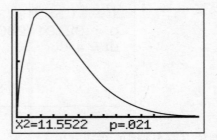

X2=11.5522   p=.021

The chi-square test can also be performed on a computer. It is important to be able to read and interpret the computer output. Given below is a computer printout of the previous example.

Chi-Square Test

Expected counts are printed below observed counts

	Creamy	Crispy	Chewy	Total	
Elemen	33	14	19	66	← observed counts
	25.67	15.40	24.93		← expected counts
Middle	21	16	17	54	
	21.00	12.60	20.40		
High	16	12	32	60	
	23.33	14.00	22.67		
Total	70	42	68	180	

Chi-Sq =   2.095 +   0.127 +   1.412 +
            0.000 +   0.917 +   0.567 +   Chi-square
            2.305 +   0.286 +   3.843 = (11.552)   test statistic

(DF = 4), (P-value = 0.021)
degrees of   P-value
freedom

The observed and expected counts are given one above the other; the individual components, $\dfrac{(\text{observed} - \text{expected})^2}{\text{expected}}$, of the chi-square test statistic are provided and summed; and the *P*-value along with the appropriate number of degrees of freedom is provided.

Chi-Square Test of Independence

The chi-square test of independence is used to compare the distributions of sample counts of two categorical variables from a single population to see if there is an association between the variables. For instance, parents of incoming freshmen in a large school district were asked if they supported school uniforms. Parents were classified by whether or not they favored uniforms, and by the type of uniform policy at their child's previous school: uniform mandatory, uniform optional, or no uniform.

The previous situation calls for a chi-square test of independence. There are two variables (uniform preference and previous school policy) with multiple categories (favorable/unfavorable for preference and mandatory/optional/no policy for previous policy) and one population under consideration (parents of incoming freshmen). One will compare the counts of the intersection of the categories of each variable to what would be expected if the variables were independent.

To perform a chi-square test of independence, use the same four-step inference procedure.

Step 1: State the parameter of interest and a correct pair of hypotheses. The null hypothesis is that the variables under consideration are independent. That is, knowing what category of one variable an individual respondent or subject falls into does not affect the probability that the individual will fall into any category of the second variable.

H_0: Variable A and Variable B are independent.

The alternative hypothesis is that knowing what category of one variable an individual respondent or subject falls into changes the probability that the individual will fall into any category of the second variable.

H_a: Variable A and Variable B are not independent.

There is no parameter of interest. One wants to know if there is an association between the two categorical variables.

Step 2: Name the inference procedure and check assumptions/conditions. The procedure is the chi-square test of independence.

The assumptions/conditions for the chi-square test for independence are the same as the other two χ^2-tests.

The expected cell counts for each cell are equal to $\dfrac{\text{(row total)}\text{(column total)}}{\text{grand total}}$.

This is equivalent to the count of the total sample that would occur given independence between variables A and B, that is, $P(A) \times P(B) = P(A \text{ and } B)$

Step 3: Calculate the test statistic and *P*-value. The chi-square test statistic is computed as

$$\chi^2 = \sum_{\substack{\text{all} \\ \text{cells}}} \frac{(\text{observed} - \text{expected})^2}{\text{expected}}.$$

The number of degrees of freedom for the chi-square test for independence is equal to (number of rows $-$ 1)(number of columns $-$ 1).

The *P*-value for the test would be the probability of obtaining a χ^2-statistic at least as large as that computed above if the null hypothesis were true. It is determined by consulting Table C, the table of χ^2-critical values, or by using technology.

Step 4: Interpret the results in context. Make a conclusion as to whether to reject, or fail to reject, the null hypothesis, based on the meaning of the *P*-value. Describe in the context of the situation what the conclusion means.

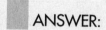 **EXAMPLE:** A random sample of parents of incoming freshmen in a large school district were asked if they supported school uniforms. Parents were classified by whether or not they favored uniforms, and by the type of uniform policy at their child's previous school: uniform mandatory, uniform optional, or no uniform. The results of the survey are shown in the table below.

		Favor Uniforms		
		Yes	No	Total
Previous Policy	Mandatory	28	7	35
	Optional	12	8	20
	No Policy	38	27	65
	Total	78	42	120

Is there evidence of a relationship between a parent's favoring uniforms and the uniform policy at the child's previous school?

ANSWER:

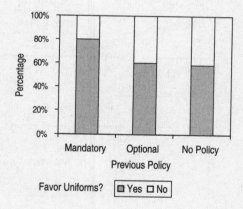

The graph suggests that there may be an association between the uniform policy at an incoming freshman's previous school and a parent's favoring uniforms. Parents of children who come from uniform-mandatory schools appear to be more favorable toward uniforms than parents of children who attended uniform-optional or no policy schools. Because the observed difference may be due to sampling variability, an inference procedure is performed to assess the chance of this.

Step 1: State the parameter of interest and a correct pair of hypotheses.

H_0: A parent's uniform favorability is independent of the uniform policy at the child's previous school.

H_a: A parent's uniform favorability is not independent of the uniform policy at the child's previous school.

Step 2: Name the inference procedure and check assumptions/conditions. Chi-square test of independence to determine if parents' favorability toward uniforms is independent of the uniform policy at the child's previous school.

(1) The data come from a simple random sample: OK—the question states that the parents were randomly selected.

(2) The size of the sample is no more than 10% of the population size: OK so long as there are at least 1,200 parents of incoming freshman in this school district. This would seem reasonable for large districts.

(3) All expected cell counts are at least 5. OK—expected cell counts are given below and all are at least 5.

Actual Counts		Favor Uniforms		
		Yes	No	Total
Previous Policy	Mandatory	28	7	35
	Optional	12	8	20
	No Policy	38	27	65
	Total	78	42	120

Expected Counts		Favor Uniforms		
		Yes	No	Total
Previous Policy	Mandatory	$\frac{(35)(78)}{120}$ $=22.75$	$\frac{(35)(42)}{120}$ $=12.25$	35
	Optional	$\frac{(20)(78)}{120}$ $=13$	$\frac{(20)(42)}{120}$ $=7$	20
	No Policy	$\frac{(65)(78)}{120}$ $=42.25$	$\frac{(65)(42)}{120}$ $=22.75$	65
	Total	78	42	120

$$\text{Expected cell count} = \frac{(\text{row total})(\text{column total})}{\text{grand total}}.$$

Note: The row and column totals for the actual and expected cell counts are equal.

Step 3: Calculate the test statistic and *P*-value.

$$\chi^2 = \sum_{\substack{\text{all} \\ \text{cells}}} \frac{(\text{observed} - \text{expected})^2}{\text{expected}}$$

$$\approx \frac{(28 - 22.75)^2}{22.75} + \frac{(7 - 12.25)^2}{12.25} + \frac{(12 - 13)^2}{13} + \frac{(8 - 7)^2}{7} + \frac{(38 - 42.25)^2}{42.25} + \frac{(27 - 22.75)^2}{22.75}$$

$$\approx 1.212 + 2.25 + 0.077 + 0.143 + 0.428 + 0.798$$

$$\approx 4.903.$$

$df = (\text{number of rows} - 1)(\text{number of columns} - 1) = (3 - 1)(2 - 1) = 2.$

P-value from Table C: the χ^2-statistic of 4.903 lies between 4.61 and 5.99 in the 2 *df* row, so $0.05 < p < 0.10$.

Step 4: Interpret the results in context. The *P*-value is between 0.10 and 0.05. If a parent's favorability toward uniforms were indeed independent of the uniform policy at a child's previous school, a chi-square statistic of 4.903 or larger would occur somewhere between 5% and 10% of the time—between 1 chance in 10 and 1 chance in 20. This is fairly likely to occur by chance and leads one to fail to reject the null hypothesis. There is insufficient evidence that a parent's favorability toward uniforms is dependent on the uniform policy at a child's previous school.

Calculator screens are shown below for the observed counts, χ^2-test, and expected counts.

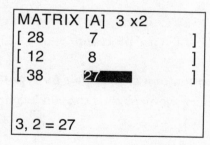

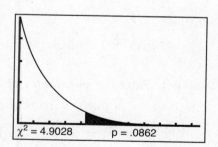

Hypothesis Tests for the Slope of the Regression Line

The hypothesis test for slope allows us to determine if there is a useful linear relationship between x and y in the population. That is, does the slope of the population model differ from zero—does y tend to change linearly with changes in x? If there is a linear relationship between the two variables, the slope should not equal zero. The null hypothesis will state that the slope of the true regression line is equal to zero. The alternative hypothesis will be either greater than, less than, or not equal zero depending on the direction in which the researcher is seeking evidence.

The formula to compute the test statistic for the population slope β_1, with the null hypothesis $H_0: \beta_1 = 0$, is $t = \dfrac{b_1}{SE_{b_1}}$, where b_1 is the slope of the sample regression line and SE_{b_1} is the standard error of the slope. The number of degrees of freedom is $n - 2$. The P-value will be computed using the t-distribution.

The assumptions/conditions for a hypothesis for a population slope are the same as for confidence intervals.

EXAMPLE: Let us revisit the height and shoe size problem from Section A. You will find the raw data there, along with a scatterplot, residual plot, and residual histogram. Is there a significant straight-line relationship between the heights of women and their shoe sizes?

ANSWER: Follow the four-step inference procedure.

Step 1: State the parameter of interest and a correct pair of hypotheses. Let β_1 be the slope of the population regression line of women's shoe sizes on height.

$H_0: \beta_1 = 0$

$H_a: \beta_1 \neq 0$.

Step 2: Name the inference procedure and check assumptions/conditions. We will use a linear regression t-test for the slope of the population regression line of women's shoe sizes on height.

(1) The mean y values for all the fixed x values are related linearly by the equation $\mu_y = \beta_0 + \beta_1 x$. OK—the scatterplot is linear and the residual plot shows no U-shaped pattern.

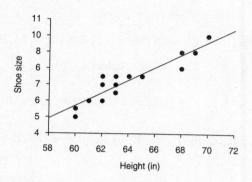 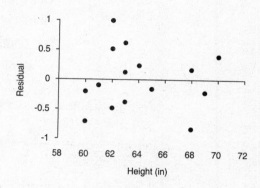

(2) For any fixed value of x, the value of each y is independent. OK—there does not appear to be any clustering in the residual plot.

(3) For any fixed value of x, the value of y is normally distributed. OK—the histogram of the residuals is symmetric and unimodal.

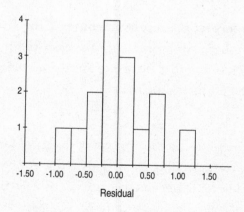

(4) For all fixed values of x, the standard deviation of y is equal. OK—the variability of the residuals across the values of x is consistent.

Step 3: Calculate the test statistic and *P*-value. Most of the time you will be given a computer printout of the regression information.

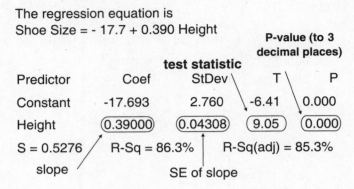

The regression equation is
Shoe Size = - 17.7 + 0.390 Height

Predictor	Coef	StDev	T	P
Constant	-17.693	2.760	-6.41	0.000
Height	0.39000	0.04308	9.05	0.000

S = 0.5276   R-Sq = 86.3%   R-Sq(adj) = 85.3%

In our sample $b_1 = 0.39$ and $SE_{b_1} = 0.043$. Substituting into the formula $t = \dfrac{b_1}{SE_{b_1}}$, we get $t = \dfrac{b_1}{SE_{b_1}} = \dfrac{0.39}{0.043} \approx 9.053$, with $n - 2 = 13$ degrees of freedom.

Using the *t*-distribution table with 13 degrees of freedom, the probability that the test statistic is 9.053 or more is less than 0.0005. Since this is a two-sided test, we double that value. The *P*-value is less than 0.001, although the very large test statistic would suggest that it is very much less.

Note that the test statistic and *P*-value (to three decimal places) are also provided on the computer printout.

Step 4: Interpret the results in context. The *P*-value calculated is very small, far less than 0.001. We have strong evidence to reject the null hypothesis in favor of the alternative, and we determine that there is a linear (or straight-line) relationship between the heights of women and their shoe sizes.

Note: Returning to the concept of duality of hypothesis tests and confidence intervals, recall that in the corresponding example in Section A we were 95% confident that the slope of the true regression line for height and shoe size was between 0.2969 and 0.4831. Zero is not in the interval, so we could conclude, at least at the $\alpha = 0.05$ level of significance, that the slope is not equal to 0.

TEST TIP

The AP Program puts about 10 years of previous free-response questions from actual exams on its website, along with scoring guidelines and, sometimes, sample student responses with the scores they received. Combined with the questions in this book, these sample questions can help you know what to expect on Section II of your AP Statistics Exam.

Calculator Tip:

Hypothesis tests for the slope of the population model can be performed with the graphing calculator. You must, however, have the raw data in lists.

In this case, the women's heights are in list L1 and shoe sizes are in list L2. From the home screen, press STAT, arrow right to TESTS, and then choose F:LinRegTTest....

```
EDIT  CALC  TESTS
B↑ 2-PropZInt...
C : χ²-Test...
D : χ² GOF-Test...
E : 2-SampFTest...
F : LinRegTTest...
G : LinRegTInt...
A : ANOVA (
```

Input the data lists, the appropriate alternative hypothesis, and where the regression equation is to be stored.

```
LinRegTTest
  Xlist : L1
  Ylist : L2
  Freq : 1
  β & ρ : ≠0 <0 >0
  ReqEQ : Y1
  Calculate
```

Arrow down to Calculate, and then press ENTER. The alternative hypothesis is shown along with the test statistic, P-value, degrees of freedom, and intercept. Scrolling down will give the slope, standard error about the regression line, r, and r^2.

```
LinRegTTest
  y=a+bx
  β≠0 and ρ≠0
  t=9.053731821
  p = 5.628817 E−7
  df = 13
↓a = −17.69333333
```

```
LinRegTTest
  y=a+bx
  β≠0 and ρ≠0
↑b = .39
  s = .5275730597
  r² = .8631147541
  r = .9290396946
```

Note the very small P-value of 0.000000563. Also, the alternative hypothesis is given in two forms. The second is $\rho \neq 0$. Rho (ρ) is the population correlation coefficient. If the slope is zero, so must be the correlation coefficient.

Time for a quiz
- Review strategies in Chapter 2
- Take Quizzes 7 and 8 at the REA Study Center
  (www.rea.com/studycenter)

Take Mini-Test 2
on Chapters 5–6
Go to the REA Study Center
(www.rea.com/studycenter)

Practice Exam

Also available at the REA Study Center (*www.rea.com/studycenter*)

Practice Exam
Section I

(Answer sheets appear in the back of the book.)

TIME: 1 hour and 30 minutes
Number of Questions—40
Percent of Total Grade—50

Directions: Solve each of the following problems, using the available space for scratchwork. Decide which is the best of the choices given and fill in the corresponding oval on the answer sheet. No credit will be given for anything written in the test book. Do not spend too much time on any one problem.

1. A study was conducted of the relationship between the number of *hours* of television a student watched in the 24-hour period before a statistics examination and the *score* on the exam. The following is a computer printout from a least-squares regression analysis.

Predictor	Coeff	StDev	T	P
Constant	93.052	3.426	27.16	0.000
Hours	−3.2319	0.7819	−4.13	0.001
$s = 7.843$	R–Sq = 55.0%		R–Sq(adj) = 51.7%	

Which of the following gives the correct value and interpretation of the correlation coefficient for the linear relationship between the test score and the number of hours of television watched?

(A) Correlation $= -0.742$. The linear relationship between the test score and the number of hours of TV watched is moderate and negative.

(B) Correlation $= 0.550$. Fifty-five percent of the variation in test scores is explained by the number of hours of TV watched.

(C) Correlation $= 7.416$. There is a relatively weak linear relationship between the test score and the number of hours of TV watched.

(D) Correlation $= 0.742$. About 74% of the data points lie on the least-squares regression line.

(E) Correlation $= -0.742$. For every additional hour of television watched, the average test score dropped by about three-fourths of a point.

2. Allison and her twin sister Brenda both take Advanced Placement European History. Allison is in the morning class while Brenda is in the afternoon class. On the final exam, Allison received a score of 89 where the scores had a mean of 87 and a standard deviation of 3. Brenda received a score of 90 on the same test except that her class scores had a mean of 89 with a standard deviation of 5. Which statement is true concerning the scores of the sisters relative to the scores on their own classes?

(A) Brenda had a higher score on the test and therefore performed better relative to her own classmate's scores.

(B) Allison performed better relative to her own classmate's scores.

(C) Allison and Brenda performed equally as well relative to their own classmate's scores.

(D) We cannot compare their scores since they are in different classes.

(E) We do not know how many students are in each class, so any comparison may not be fair.

3. The maker of printer cartridges for laser printers wants to estimate the mean number of documents μ that can be printed on a new high-speed printer. The company decides to test the cartridge on two dozen different laser printers. Each document is identical in number of words and amount of graphics. A histogram of the number of pages printed for each printer shows no outliers and is fairly bell-shaped. The mean and standard deviation of the sample were 3,875 sheets, and 170 sheets, respectively. It can be assumed that the laser printers were a random sample of all laser printers on the market. Which of the following is the correct formula for a 90% confidence interval for the mean number of pages printed with the new type of cartridge?

(A) $3,875 \pm 1.711 \times \dfrac{170}{\sqrt{24}}$

(B) $3,875 \pm 1.714 \times \dfrac{170}{\sqrt{24}}$

(C) $3,875 \pm 1.711 \times \dfrac{170}{\sqrt{23}}$

(D) $3,875 \pm 1.714 \times \dfrac{170}{\sqrt{23}}$

(E) The company should only compute a 95% confidence interval for these data.

4. The manager at an employment agency is interested in knowing if there is a significant difference in the mean ages of executives at two rival computer software companies. This information will help him to best place people in positions at these companies. He was allowed access to the ages of all of the executives. He found a 95% *t*-confidence interval for the difference in the mean ages of all the executives at both companies. There are 86 executives at one company and 79 at the other. Why is the information provided by this interval NOT useful for this situation?

(A) There is too much variation in the length of time the workers have been at their jobs.

(B) There is too much variation in the ages of the executives at both companies.

(C) Both populations of executive ages were used, so a confidence interval to estimate the difference in mean ages is not necessary.

(D) The sample sizes are not the same.

(E) The shapes of the distribution of ages are probably not normal.

5. The owner of a camera shop deals with, on average, 40 customers per day. Typically, 15% of these customers require repairs to camera equipment. If *X* is the number of customers the owner sees on a given day until one needs repairs, which of the following best describes the probability distribution of *X*?

(A) Binomial

(B) Chi-square

(C) Geometric

(D) Normal

(E) *t*

6. A well-known diet doctor has developed a new formula for a drug that will suppress appetites. The Food and Drug Administration (FDA) wants this doctor to submit research to show that there is a significant difference in weight loss for patients using the new drug versus the old drug. Which of the following would be the best method to obtain results the FDA is requesting?

    (A) The doctor should randomly choose patients and allow them to select which drug they want to use.

    (B) The doctor should randomly choose patients to take the new drug and ask patients using the old drug how much weight they have lost.

    (C) The doctor should randomly assign patients to two groups: one group takes the new drug and the other group takes the old drug.

    (D) The doctor should randomly assign the patients to two groups: one group takes the new drug and the other group takes a placebo.

    (E) The doctor assigns the new drug to patients who have more than 10 pounds to lose and the old drug to patients with less than 10 pounds to lose.

7. A least-squares regression model of y on x is $\hat{y} = -73.71 + 27.76x$. The residual graph is shown below.

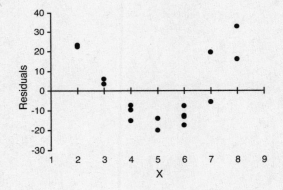

A transformation is made and the residual plot for the model $\widehat{\log y} = -0.1348 + 2.623 \log x$ is shown.

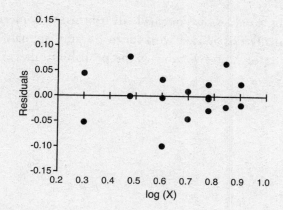

Which of the following is the correct conclusion given the graphs above?

(A) Neither model provides an appropriate relationship between the variables.

(B) There is a linear relationship between the two variables; the first model is more appropriate.

(C) There is a linear relationship between the two variables; the second model is more appropriate.

(D) There is a nonlinear relationship between the two variables; the first model is more appropriate.

(E) There is a nonlinear relationship between the two variables; the second model is more appropriate.

8. In a test of H_0: $\mu_1 = \mu_2$ versus H_A: $\mu_1 > \mu_2$, samples from approximately normal populations produce means of $\bar{x}_1 = 24.3$ and $\bar{x}_2 = 22.4$. The t-test statistic is 1.44 and the P-value is 0.085. Based upon these results, which of the following conclusions can be drawn?

(A) There is good reason to conclude that $\mu_1 > \mu_2$ because $\bar{x}_1 > \bar{x}_2$.

(B) About 8.5% of the time $\bar{x}_1 = \bar{x}_2$.

(C) About 8.5% of the time $\mu_1 > \mu_2$.

(D) The null hypothesis can be rejected since $\bar{x}_1 - \bar{x}_2$ is greater than the t-test statistic.

(E) The observed difference between $\bar{x}_1$ and $\bar{x}_2$ is significant at the $\alpha = 0.10$ level.

9. X and Y are independent random variables. X is normally distributed with mean 100 and standard deviation 8. Y is normally distributed with mean 96 and standard deviation 6. For randomly generated values of X and Y, what is the probability that X is greater than Y?

(A) 0.0401

(B) 0.6554

(C) 0.8273

(D) 0.9772

(E) 1

10. A student studying the commuting habits of students at a particular college wants to compare the mean commuting time for undergraduates with that of graduate students. The question is whether or not there is a significant difference between the mean travel time. Random samples of size 50 are taken from each group and their commuting times recorded. Which would be the appropriate test procedure to use in this situation?

(A) One-sample z-test

(B) Two-sample z-test

(C) Paired t-test

(D) Two-sample t-test

(E) Chi-square test of homogeneity

11. The salaries for an electronics company were posted in their annual sales report for all stockholders. The president of the company makes the most money. His salary was mistakenly shown to be $25,000 less than it actually is. Even with the mistake, his salary is still higher than anyone else's. Which statistic did NOT change after the mistake was corrected?

(A) Standard deviation

(B) Mean

(C) Median

(D) Variance

(E) Range

12. Which of the following allow inference about a population parameter?

   I. A randomized controlled experiment using volunteers

   II. An observational study using volunteers

   III. A survey of a random sample from the population

   (A) I only

   (B) II only

   (C) III only

   (D) I and II only

   (E) I and III only

13. The least-squares regression line has been computed to predict the yield of a certain variety of roses from the number of seeds planted. The equation is $\hat{y} = -1.05 + 0.385x$. What does the model predict the yield will be for 18 seeds planted?

   (A) About 6

   (B) About 18

   (C) About 50

   (D) About 690

   (E) There are not enough seeds to yield any plants

14. Baccarat is a casino card game between a "player" and a "dealer" where bettors wager on which will have the higher hand. Let Y represent the amount won or lost by the bettor on a single wager on the "dealer." The expected value of Y is $-\$1.06$ and the standard deviation of Y is $\$92.72$. If a bettor places 400 wagers on the "dealer" during the course of a gambling session, what is the approximate probability that the bettor ends up with a positive outcome, that is, makes money?

   (A) 0

   (B) 0.410

   (C) 0.495

   (D) 0.500

   (E) 0.590

15. There is presently a dispute about allocation of federal funds to schools in two regions of a large school district. The director of federal funds, whose spouse works in the West region, has been accused of providing more money to schools in the West region than the East region.

    Two sets of data are gathered by the school board: one listing the amount of money allocated per pupil in the 26 West region schools, the other listing the same for the 26 East region schools. The school board wants to make a graph showing a comparison of the spending between the regions. Which of the following graphs is *inappropriate* to make such a comparison?

    (A) A scatterplot

    (B) Parallel boxplots

    (C) Back-to-back stemplots

    (D) Parallel dotplots using the same scale

    (E) Parallel histograms using the same scale

16. The number of team flags the booster club will sell at a football game has the probability distribution as shown in the table below.

Number of team flags (X)	5	6	7	8	9	10
P(X)	0.20	0.15	0.10	0.25	0.18	0.12

    If each team flag costs $8.00, what is the expected amount of money the booster club will take in at a football game?

    (A) $7.42

    (B) $8.00

    (C) $53.15

    (D) $59.36

    (E) $64.00

17. In the computation of a confidence interval, if the sample size is not changed but the confidence level is changed from 99% to 95%, you can expect

    (A) an interval with the same width since the mean has not changed

    (B) an interval with the same width since the sample size has not changed

    (C) an interval that is wider

    (D) an interval that is narrower

    (E) The change cannot be determined from the information given

18. The makers of Save-More Showerheads claim that their showerhead will save water and therefore save money on water bills. They cite evidence from a recent study where sales records from a home improvement center were used to identify customers who purchased a Save-More Showerhead. Twenty of these customers were contacted and 19 of them indicated that they used less water in the month following the installation of the showerhead. Which of the following statements best describes the claim made by the makers of the showerheads?

(A) It is valid. The evidence shows lower water use by nearly all customers

(B) It is invalid because there were not enough customers in the study.

(C) It is invalid because changes in water usage due to the showerhead are confounded with other variables.

(D) It is invalid because Save-More should have sold more than one type of showerhead.

(E) It is invalid because no other brands of showerheads were included in the study.

19. A significance test is performed with a significance level of $\alpha = 0.05$. For a particular value of the population parameter, the probability of committing a Type II error is computed to be $\beta = 0.13$. What is the power of the test for this situation?

(A) 0.05

(B) 0.13

(C) 0.82

(D) 0.87

(E) 0.95

20. Alex, Bryan, and Charlie are all playing tennis matches in a tournament against different opponents. Based on previous performances, there is a 0.4 probability that Alex will win his first match, a 0.3 probability that Bryan will win his first match, and a 0.2 probability that Charlie will win his first match. If the chance that each wins his match is independent of the others, what is the probability that none of them wins in their first matches?

(A) 0.024

(B) 0.304

(C) 0.336

(D) 0.700

(E) 0.900

21. A study was conducted to determine if the proportion of cars in a given area that do not meet exhaust emission standards has changed from 10 years ago when it was 8%. A test of H_0: $p = 0.08$ against H_A: $p \neq 0.08$ was performed with a significance level of $\alpha = 0.05$. Given that the data showed that there was a significant difference in the proportion of cars not meeting emission standards now versus 10 years ago, which of the following statements is true?

    (A) The researchers would have reached the same conclusion at $\alpha = 0.01$ and $\alpha = 0.10$.

    (B) The researchers would have reached the same conclusion at $\alpha = 0.01$, but not at $\alpha = 0.10$.

    (C) A 95% confidence interval for the population proportion p would contain 0.08.

    (D) A 90% confidence interval for the population proportion p would contain 0.08.

    (E) The correct statement cannot be determined without knowing the P-value.

22. An entomologist is studying fruit flies to determine if eye color is a sex-linked trait. The hypothesis is that a particular generation of flies will be evenly divided between males and females, with eye color—red or white—evenly divided within each gender. The actual results are shown below.

Male, red eyes	Male, white eyes	Female, red eyes	Female, white eyes
19	22	33	26

    What is the value of the χ^2-statistic for a goodness-of-fit test on these data?

    (A) 0

    (B) 0.17

    (C) 4.28

    (D) 4.40

    (E) 27.5

23. A pediatrician is looking over records of female patients, specifically girls aged 10 years. The tallest 10% of girls had heights of 146.0 cm or more, while the shortest 25% of girls had heights of 133.0 cm or less. If the heights of 10-year-old girls are approximately normally distributed, what are the mean and standard deviation of the heights?

    (A) Mean = 136.7 cm; standard deviation = 7.27 cm

    (B) Mean = 137.5 cm; standard deviation = 6.64 cm

    (C) Mean = 139.5 cm; standard deviation = 5.08 cm

    (D) Mean = 139.5 cm; standard deviation = 9.70 cm

    (E) Mean = 141.5 cm; standard deviation = 3.52 cm

24. The display below shows the cumulative relative frequency histogram of scores from the 20-question math placement examination taken by 40 freshmen upon entering a high school.

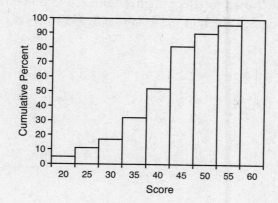

Which of the following is a correct statement based on the information in the display?

(A) The median score is 30.

(B) Most students scored above 50.

(C) No one scored 35 on this test.

(D) About four times as many students scored 30 than 20.

(E) There were about equal numbers of students with scores between 50 and 60.

25. What is the major difference between an experiment and an observational study?

(A) A treatment is imposed in an experiment.

(B) An observational study can establish cause–effect relationships.

(C) There are two control groups instead on one in an experiment.

(D) Observational studies use only one population.

(E) Experiments are blinded.

26. Five homes from a subdivision will be randomly selected to receive 1 month of free cable TV. There are 80 homes in the subdivision. The homes are assigned numbers 01–80 and the random number table below is used to select the five homes. No home may receive more than one free month of service. Which of the following is a correct selection of the five homes?

$$99154 \quad 70392 \quad 23889 \quad 92335$$
$$92210 \quad 70439 \quad 08629 \quad 73299$$

(A) 9, 1, 5, 4, 7

(B) 15, 47, 03, 23, 23

(C) 15, 47, 03, 23, 35

(D) 99, 70, 23, 92, 08

(E) 99, 15, 47, 03, 92

27. A tire manufacturer claims that the average tire tread will last about 50,000 miles. An automobile magazine does not believe this claim; they believe the tire tread to wear out sooner. If μ represents the true number of miles the tread will last, which of the following pairs of hypotheses is correct to test the claim?

(A) $H_0 : \mu = 50,000$
$H_a : \mu > 50,000$

(B) $H_0 : \mu = 50,000$
$H_a : \mu < 50,000$

(C) $H_0 : \mu = 50,000$
$H_a : \mu \neq 50,000$

(D) $H_0 : \mu < 50,000$
$H_a : \mu \geq 50,000$

(E) $H_0 : \mu > 50,000$
$H_a : \mu \leq 50,000$

28. A fair six-sided die has four faces painted green and two faces painted red. The die is rolled 12 times; let X be the number of times a red face shows. This procedure of 12 rolls is repeated 100 times. Which of the following plots is most likely to display the distribution of X from these 100 trials?

(A)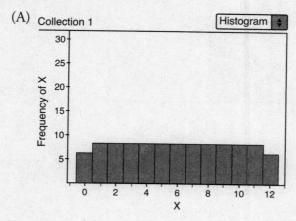

(B)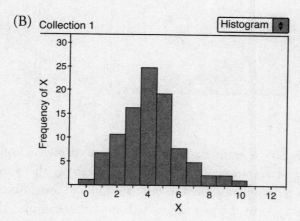

(C)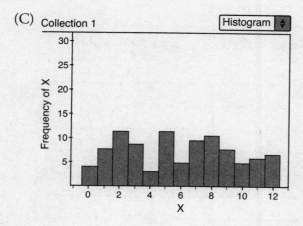

(D)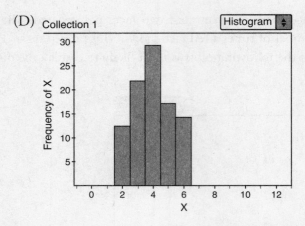

(E)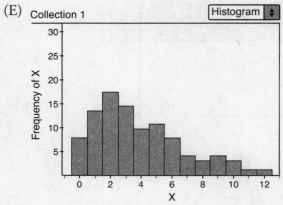

29. A group of 10 trained volunteers is taste-testing a new cola for a soft drink manufacturer. There are two formulas under consideration. One of the characteristics the tasters will rate is "spice." Each volunteer is randomly assigned to taste a sample of one formula, rate it on a scale of 1 to 10, and then repeat the procedure with the other sample. The data are given in the table below.

Tester	Spice for Formula A	Spice for Formula B
1	9	6
2	7	5
3	8	8
4	6	6
5	9	7
6	7	7
7	7	8
8	9	7
9	4	5
10	7	6

What is the number of degrees of freedom associated with the appropriate t-confidence interval to see which formula has a higher "spice" rating?

(A) 9

(B) 10

(C) 16

(D) 18

(E) 20

30. The cafeteria manager at a university wants to conduct a survey of 100 students about the quality of food for students who live in the dorms. Which of the following will give the best representation of all dorm students?

   (A) Survey every eighth student who arrives in the cafeteria.

   (B) Survey the first 100 students who arrive at the cafeteria at dinner time.

   (C) Select a random sample of dorm students.

   (D) Hand out a survey to any student who will take the time to answer the questions.

   (E) Put surveys in the dormitory mailboxes.

31. The parallel boxplots below represent the amount of money collected (in dollars) in a 1-day fundraiser from each of 16 boys and 16 girls in a certain neighborhood in town. Which of the following is known to be true about the data by looking at the plots?

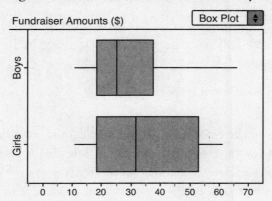

   (A) The boys collected more money than the girls.

   (B) The median of the amounts the girls collected is larger than the median of the amounts the boys collected.

   (C) The distribution of the amounts the boys collected is fairly symmetric.

   (D) The interquartile range of both distributions is the same.

   (E) There is less variability in the amounts the girls collected than the boys.

32. In which of the following instances would the t-distribution be used to model the sampling distribution of a sample statistic instead of the z-distribution?

    (A) Constructing a confidence interval for the population proportion; the population proportion, p, is unknown.

    (B) Conducting a matched-pairs hypothesis test for the population mean difference; the population standard deviation of the difference, σ_d, is known.

    (C) Constructing a confidence interval for the difference between two population means; the population standard deviations, σ_1 and σ_2, are unknown.

    (D) Conducting a hypothesis test for the difference between two population proportions; the population proportions, p_1 and p_2, are unknown.

    (E) Constructing a confidence interval for a population mean; the population standard deviation, σ, is known.

33. A poll was conducted to find out the percentage of town residents that would like an additional traffic light on Main Street. Of the 257 respondents, 61% said they would like the new traffic light. The poll had a margin of error of 3%. Which of the following correctly describes the margin of error?

    (A) There are more than 3% of the town residents that want the traffic light.

    (B) There are less than 3% of the town residents that want the traffic light.

    (C) There is most likely more than 3% difference between the percentage in the sample and the percentage of the population.

    (D) There is most likely less than 3% difference between the percentage in the sample and the percentage of the population.

    (E) About 3% of the town residents did not respond to the poll.

34. A linear relationship exists between the amount of money spent (in thousands of dollars) on advertising and the amount (in thousands of dollars) of sales for a particular shoe manufacturer. The least-squares regression line was calculated to be $\hat{y} = 25.2 + 6.2x$, where x is the money spent on advertising and $\hat{y}$ is the amount in sales. What is the estimated increase in sales (in thousands of dollars) for every $3,000 spent on advertising?

    (A) 6.2

    (B) 18.6

    (C) 21.5

    (D) 25.2

    (E) 43.8

35. The manufacturer of an automobile battery claims that the battery provides at least 900 cold cranking amps (CCA) of current to a car's starter. To maintain this claim, a quality control inspector takes a random sample of 12 batteries each hour and measures the CCA put out by each battery. The inspector will shut down the assembly process if there is evidence that the mean CCA of the 12 batteries has dropped below 900.

Explain the result of a Type I error in this situation.

(A) The manufacturer will decide that the mean battery CCA is greater than 900 CCA when in fact it is 900 CCA.

(B) The manufacturer will decide that the mean battery CCA is greater than 900 CCA when in fact it is less than 900 CCA.

(C) The manufacturer will decide that the mean battery CCA is greater than 900 CCA when in fact it is greater than 900 CCA.

(D) The manufacturer will decide that the mean battery CCA is 900 CCA when in fact it is greater than 900 CCA.

(E) The manufacturer will decide that the mean battery CCA is less than 900 CCA when in fact it is at least 900 CCA.

36. The school Parent–Teacher Association (PTA) wants to sample parents about the type of school lunches they prefer for their children. At a recent Open House parent night, the volunteer conducting the survey decided to hand the questions out to every 10th parent that came to the Open House. This sampling method is called a

(A) convenience sample

(B) multistage sample

(C) cluster sample

(D) stratified random sample

(E) simple random sample

37. Which of the following residual plots and corresponding residual histograms would indicate that the conditions for inference for the slope of the least-squares regression line have been satisfied?

(A)

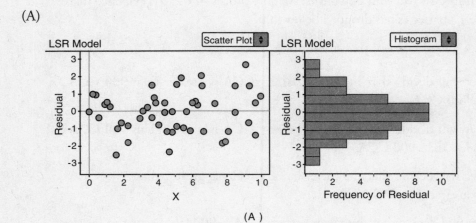

(A)

(B)

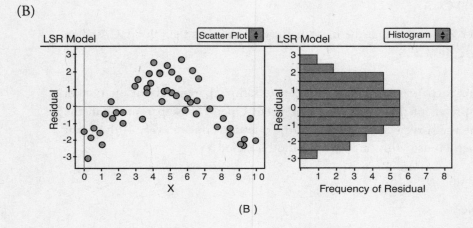

(B)

(C)

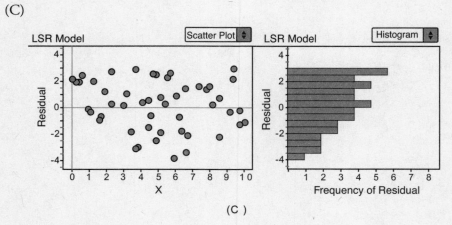

(C)

(D)

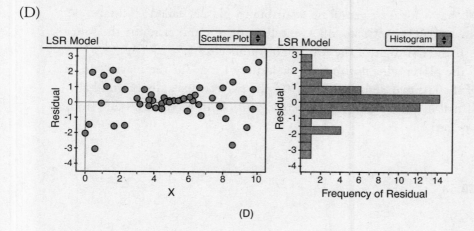

(D)

(E)

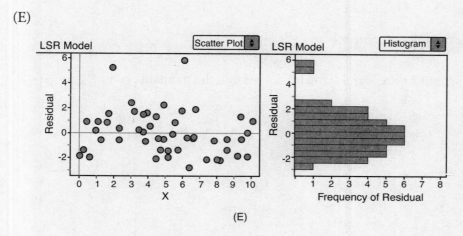

(E)

38. Researchers from a tire manufacturer want to conduct an experiment to compare tread wear of a new type of tires with the old design. They are going to secure 200 past customers as volunteers to participate in the experiment for 6 months. Tires will be placed on the front two wheels of the cars. The amount of tread wear will be measured after the tires are driven for a designated amount of time. The volunteers will not be able to distinguish between the types of tires on their vehicle. Of the following, which is the best design for this experiment?

(A) Since customers are volunteering their vehicles for the study, let each choose which type of tire they want on their car.

(B) Randomly assign the new type of tire to both wheels on 100 cars; the other 100 cars receive the old type of tire on both wheels.

(C) Put a new type of tire on the left wheel of each car and the old type of tire on the right wheel.

(D) Assign each car the new tire to both wheels. After 3 months, replace them with the old tire for another 3 months.

(E) Randomly assign one of the types of tires to both wheels. After 3 months, replace them with the other type of tire for another 3 months.

39. A particular golf club has a form of gambling available to players called "Pulltabs." A player pays \$1 to pull a small paper ticket off a spindle, open it, and examine the four-digit number inside. If the last digit of the four-digit number is "1," "2," "3," "4," "5," "6," "7," "8," or "9," the player wins nothing. If the last digit is a "0," the player wins a golf ball. However, if the last two digits are both "0," the player wins a dozen golf balls. If the value of a golf ball is \$2, what is the expected gain/loss in monetary value from buying one Pulltabs ticket?

(A) A loss of about \$0.48

(B) A loss of about \$0.58

(C) A loss of about \$0.99

(D) A gain of about \$0.42

(E) Approximately breakeven—no gain or loss

40. For a simple random sample of size n and standard deviation s, the standard error of $\bar{x}$ is

(A) $\dfrac{s}{\sqrt{n}}$

(B) $\dfrac{\sigma}{n}$

(C) s

(D) σ

(E) $\dfrac{s}{\sqrt{n-1}}$

Section II

Part A

TIME: Questions 1–5

Spend about 65 minutes on this part of the exam.

Percent of Section II grade—75

Directions: Show all your work. Indicate clearly the methods you use, because you will be graded on the correctness of your methods as well as on the accuracy of your results and explanation.

1. The table below shows science test scores obtained by groups of fourth graders in different classrooms.

   Room A: 65 73 74 75 76 77 79 81 81 82 83 83 84 84 85 88 90 90 92 96

   Room B: 60 64 64 64 65 66 67 68 72 72 73 74 74 76 77 79 80 81 81 83

   Room C: 69 69 73 73 74 75 76 77 78 79 80 81 82 83 85 88 88 89 90 92

   (a) Create three parallel box-and-whisker plots using the same scale. Be sure to label the room letter for each plot.

   (b) Compare the three distributions.

2. The American Red Cross is responsible for collecting blood donations across the United States. The blood it collects can be divided into four types based on its antibodies: A, B, AB, and O. Blood can also be divided into two types by Rh-factor: positive and negative. The most sought-after blood type is O-negative, since it can be given to a person with any blood type or Rh-factor.

   Forty-five percent of Americans have O-type blood; 16% of Americans are Rh-negative.

   (a) If blood type and Rh-factor are independent, what percentage of Americans have O-negative blood?

   (b) The actual percentage of Americans with O-negative blood is 6%. If an American is selected at random, what is the probability that they are type O given that they are Rh-negative?

   (c) If Americans go to a blood bank to donate blood in a manner that is essentially random, what is the mean number of people that will donate blood before someone with O-negative is seen?

3. In a study designed to investigate the relationship between mathematical reasoning of children aged 10 to 19, a single spatial reasoning test was administered to groups of 25 randomly selected children at each age level. The table below shows the mean scores of the 25 students in each group.

Relationship between Age and Spatial Reasoning Test Score

Group	1	2	3	4	5	6	7	8	9	10
Age	10	11	12	13	14	15	16	17	18	19
Mean Scores	65	63	70	73	81	79	81	80	83	85

Summary Statistics	Mean Age	SD Age	Mean Scores	SD Mean Scores	Correlation
	14.5	3.03	76	7.75	0.929

(a) Find the equation of the least-squares regression line.

(b) Calculate *and* interpret the value of r^2.

(c) The plot of the residuals is shown below. Comment on the appropriateness of the linear model.

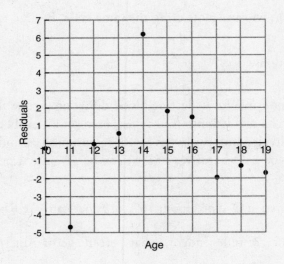

4. Worldwide Airlines is studying the weights of passengers' carry-on bags. The airline is interested in how the weights of carry-on bags of pleasure travelers compare with those of business travelers.

Passengers with carry-on bags were randomly selected over a period of several weeks. Selected passengers were asked if their trip was for business or pleasure, and total weight of their carry-on bags measured. The summary statistics are shown in the table below.

Descriptive Statistics

Variable Type	N	Mean	Median	TrMean	StDev
Weights Business	31	33.68	35.00	33.82	4.62
Weights Pleasure	59	8.92	6.00	8.62	7.42

Variable Type	SE Mean	Min	Max	Q1	Q3
Weights Business	0.830	25.0	41.0	30.0	38.0
Weights Pleasure	0.966	0.00	23.0	2.00	15.0

(a) Compute and interpret the 95% confidence interval for the true difference in average weights of carry-on bags between business and pleasure passengers.

(b) Dotplots of the raw data are provided. Address whether all conditions for the inference procedure from part (a) have been met.

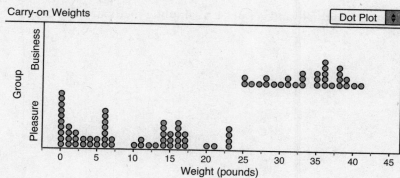

5. The administrators in a high school are thinking of changing the school's parking policy effective about 3 weeks after school begins. The administration has asked the Student Council to conduct a survey during the first week of school to determine what students who own cars think about the proposal.

The student body has the following distribution. The number of students in each grade who own cars is also provided.

Grade	Freshmen	Sophomores	Juniors	Seniors
Population	500	550	500	450
Own cars	0	180	315	405

The Student Council has decided to survey 100 students. The student body president wants to conduct a simple random sample to obtain the names of the 100 students to be surveyed. The student body secretary wants to use a stratified sampling technique to obtain the names of the 100 students.

(a) If the student body president's plan is chosen, describe the procedure used to select the 100 students.

(b) If the student body secretary's plan is chosen, describe the procedure used to select the 100 students.

(c) What are the advantages and disadvantages of each sampling method?

Section II
Part B

TIME: Question 6

Spend about 25 minutes on this part of the exam.

Percent of Section II grade—25

Directions: Show all your work. Indicate clearly the methods you use, because you will be graded on the correctness of your methods as well as on the accuracy of your results and explanation.

6. The championship series for professional baseball, basketball, and hockey involve two teams playing a series in which the first team to win four games wins the series. We are interested in studying the length of series. Consider two different championship series. In the first series, the two teams are evenly matched. That is, each has an equally likely chance of winning. In the second series, one team beats the other 70% of the time.

   (a) Explain how you would conduct a simulation using a random digit table to determine the length of a series if each team is evenly matched.

   (b) Explain how you would conduct a simulation using a random digit table to determine the length of a series if one team wins 70% of games between the teams.

   (c) Using the random digit table below, conduct your simulations described in parts (a) and (b) five times each. By marking directly on or above the table, make your procedure clear enough for someone to understand. Record the final number of games played in each series.

93015	93615	03413	72295	97291
36377	14756	03768	53840	55333
63186	77781	69103	43259	01660
23825	65704	49525	84121	44856
97187	05901	61053	04173	07717

(d) The results of two simulations of 100 trials each are shown below.

| | Number of Trials | |
Length of Series	Simulation 1	Simulation 2
4 games	12	25
5 games	25	30
6 games	33	27
7 games	30	18

Identify which simulation, 1 or 2, represents series in which the teams are evenly matched. Explain your reasoning.

(e) Use the distributions in part (d) to compute an estimate of the difference in length of series between two teams that are evenly matched versus a series where one team wins 70% of the time.

Answer Key

Section I

1. (A)	11. (C)	21. (E)	31. (B)
2. (B)	12. (C)	22. (D)	32. (C)
3. (B)	13. (A)	23. (B)	33. (D)
4. (C)	14. (B)	24. (E)	34. (B)
5. (C)	15. (A)	25. (A)	35. (E)
6. (C)	16. (D)	26. (C)	36. (A)
7. (E)	17. (D)	27. (B)	37. (A)
8. (E)	18. (C)	28. (B)	38. (E)
9. (B)	19. (D)	29. (A)	39. (B)
10. (D)	20. (C)	30. (C)	40. (A)

Detailed Explanations of Answers

Section I

1. **(A)**

    The output gives us a value of R-squared at 0.550. The correlation, r, is either the positive or negative square root of this value. Correlation has the same sign as the slope of the regression line. Since the regression line has a negative slope (-3.2319), the correlation must be $-\sqrt{0.550} = -0.742$. Correlation tells us the direction and strength of a linear relationship. Correlations between 0.5 and 0.8 (or -0.5 and -0.8) are usually considered "moderate."

2. **(B)**

    We can see how one score compares to others in its distribution by looking at its z-score. Allison's z-score is $z = \dfrac{89-87}{3} = \dfrac{2}{3}$. Brenda's z-score is $z = \dfrac{90-89}{5} = \dfrac{1}{5}$. Since Allison's z-score is higher, she did better relative to the scores in her own class, even though Brenda's raw score was higher.

3. **(B)**

    A 90% confidence interval for a mean is given by $\bar{x} \pm t_{n-1}^{*} \dfrac{s}{\sqrt{n}} = 3875 \pm 1.714 \dfrac{170}{\sqrt{24}}$. Note that 1.714 is the t-score for a 90% confidence interval with $24 - 1 = 23$ degrees of freedom.

4. **(C)**

    The manager used the ages of *all* the executives at both companies, so he conducted a census. Inferential statistics use a sample to make generalizations about a population. Since we already have the entire population, there is no need for any inference!

5. **(C)**

The variable of interest in a geometric setting is the number of trials until the first success.

6. **(C)**

If the FDA wants to see if one of the drugs leads to more weight loss, a randomized experiment is necessary. Choices A, B, and E have no random assignment to groups. Choice D compares the new drug to a placebo, which is not the goal of the study.

7. **(E)**

The first residual plot clearly shows a curved pattern. This means that the linear model is not appropriate, and the relationship between the two variables is nonlinear. The second model has a residual plot with more random scatter, suggesting that the model is more appropriate.

8. **(E)**

With a P-value of 0.085, these results are significant at any alpha level greater than 0.085. Choices B and C are incorrect interpretations of a P-value. Choice D is nonsense—a difference in means and a t-test statistic measure completely different things. Choice A is wrong, because it takes more than $\bar{x}_1 > \bar{x}_2$ to conclude $\mu_1 > \mu_2$. The difference must be shown that it was unlikely to happen by chance alone.

9. **(B)**

Since X and Y are both distributed normally, $X - Y$ is distributed normally. We calculate $\mu_{X-Y} = \mu_X - \mu_Y = 100 - 96 = 4$ and $\sigma^2_{X-Y} = \sigma^2_X + \sigma^2_Y = 36 + 64 = 100$.

So $\sigma_{X-Y} = 10$. Now $P(X > Y) = P(X - Y > 0) = P\left(z > \dfrac{0-4}{10}\right) = P(z > -0.40) = 0.6554$.

10. **(D)**

We have two independent samples, so the one-sample tests are out. Scores on exams are numerical variables, so the chi-square test is out. Because we do not know the population standard deviations of the two groups, we use t-procedures instead of z-procedures.

11. **(C)**

   The median is unaffected if the largest value of a data set is changed and remains the largest. Standard deviation, mean, and variance are all calculated using every value of the dataset, so they will change. The range is calculated from the largest and smallest values, so it will change.

12. **(C)**

   We can only make inferences about a population parameter if our sample is representative of that population. Using volunteers does not give us a representative sample. We can do inference with data from volunteers in a randomized comparative experiment, but we cannot generalize the results to the population.

13. **(A)**

   The question asks us for a predicted value, so we can substitute 18 for x. $\hat{y} = -1.05 + 0.385(18) = 5.88$. Note that we are asked for the yield. Other questions (like No. 34) ask for the change or increase in the response variable.

14. **(B)**

   The probability that the sum of all 400 wagers is greater than zero is equivalent to the probability that the average of all 400 wagers is greater than zero. Since we have a large sample size, $\bar{x}$ is approximately normally distributed with a mean of -1.06 and a standard deviation of $\dfrac{92.72}{\sqrt{400}}$. So,

$$P(\bar{x} > 0) = P\left(z > \frac{0 + 1.06}{92.72 / \sqrt{400}} \right) = P(z > 0.2286) = 0.410.$$

15. **(A)**

   Scatterplots are only used with paired data. There is no pairing between individual schools in the West region and the East region. All of the other graphs can be used to compare two distributions.

16. **(D)**

   First, find the expected value of x.

   $E(x) = 5(0.20) + 6(0.15) + 7(0.10) + 8(0.25) + 9(0.18) + 10(0.12) = 7.42$. The expected amount of revenue is (7.42 flags)($8/flag) = $59.36.

17. **(D)**

   A decrease in confidence level narrows the confidence interval. We gain confidence in capturing the parameter with wider intervals. Note that z^* for a 99% CI tells us it is 2.326 standard deviations wide, but for a 95% CI is only 1.960 standard deviations wide.

18. **(C)**

The claim from this observational study is seriously flawed. This is because there are countless other confounding variables that come into play when considering reduction in water use. The most obvious one is that people who bought the showerheads were probably trying to save water to begin with, and it cannot be determined if their savings was due to the showerhead or the fact that they may be conserving water any place they can. The sample size is not too small—well-designed studies can be done with small sample sizes. Other types of showerheads, whether from the company in question or a competitor, are not the issue here.

19. **(D)**

The power of the test is the probability of rejecting a false null hypothesis. A Type II error, β, is the probability of failing to reject a false null hypothesis. So power $= 1 - \beta = 1 - 0.13 = 0.87$.

20. **(C)**

Since their match outcomes are independent of each other, we can find the probability that all three lose by simply multiplying the three individual probabilities of losing. $P(\text{Alex loses}) = 1 - 0.4 = 0.6$. $P(\text{Bryan loses}) = 1 - 0.3 = 0.7$. $P(\text{Charlie loses}) = 1 - 0.2 = 0.8$. $P(\text{All 3 lose}) = (0.6)(0.7)(0.8) = 0.336$.

21. **(E)**

Choices A and B are false. When you reject a null hypothesis, you would reject it at any significance level α greater than or equal to the one you used, but you cannot know if you would reject it at lower values of α without knowing the P-value. So rejecting at $\alpha = 0.05$ also means rejecting at $\alpha = 0.10$, but not necessarily $\alpha = 0.01$. Choice C is definitely false. A $(1 - \alpha)$ confidence interval represents a set of values for the parameter that would not have been rejected in a two-sided test of significance. Since the null hypothesis of $p = 0.08$ was rejected at $\alpha = 0.05$, then 0.08 would not be in the 95% confidence interval. Choice D is also false; if 0.08 is not in the 95% confidence interval, it cannot be in the narrower 90% confidence interval.

22. (D)

Since the hypothesis is that colors and genders are equally divided, the expected values of each cell are all 25. We need to calculate $\sum \dfrac{(\text{observed} - \text{expected})^2}{\text{expected}}$.

Observed O	Expected E	$\dfrac{(O-E)^2}{E}$
19	25	1.44
22	25	0.36
33	25	2.56
26	25	0.04
	TOTAL	4.40

23. (B)

First, find the z-scores associated with the 90th and 25th percentiles. $z_{0.90} = 1.28$ and $z_{0.25} = -0.67$. Using these z-scores, we can write the equations $1.28 = \dfrac{146 - \mu}{\sigma}$ and $-0.67 = \dfrac{133 - \mu}{\sigma}$. We can then proceed to use algebra to solve this system:

$$\sigma = \frac{146 - \mu}{1.28} \text{ and } \sigma = \frac{133 - \mu}{-0.67}$$

$$\text{so} \quad \frac{146 - \mu}{1.28} = \frac{133 - \mu}{-0.67}$$

$$-0.67(146 - \mu) = 1.28(133 - \mu)$$

$$-97.82 + 0.67\mu = 170.24 - 1.28\mu$$

$$1.95\mu = 268.06$$

$$\mu = 137.5.$$

24. (E)

The change in cumulative percentage from 45 to 50 to 55 to 60 is fairly constant, indicating that about the same number of students scored 50, 55, and 60. Choice A is incorrect. To find the median in a cumulative graph, trace a horizontal line from 50 on the y-axis. We see that here the median is a score of 40. Choice B is incorrect. The height of the bar at score 50 is a cumulative percentage of about 90. This means 90% scored below or at 50. Choice C is wrong. The bars for scores 30 and 35 are not of the same height, so some students must have scored 35. Choice D is also wrong. It appears that about 5% of students scored 20. Looking at the difference in height between the bars for scores 25 and 30, it appears that about 5% of students also scored 30.

25. **(A)**

   This is the most important difference between experiments and observational studies. Experiments have a treatment imposed but are not always blinded. Observational studies can look at more than one population, but cannot establish cause–effect relationships.

26. **(C)**

   Since homes are assigned two-digit numbers, we will take digits two at a time off the table. If the two-digit number is greater than 80, or one that has already been selected, it will be ignored. So we ignore 99, select 15, select 47, select 03, ignore 92, select 23, ignore 99, 88, and 23 again, and select 35. There are other methods that will work, but all other choices either use one-digit instead of two (A), have repeated selections (B), and/or ignore the 01–80 assignment (D and E).

27. **(B)**

   The null hypothesis is the existing claim. We always look for it to be an equation. The alternative is what someone is trying to show, in this case, the magazine's claim.

28. **(B)**

   The random variable X has a binomial distribution in this problem. The mean of the distribution of X is $\mu_X = np = (12)\left(\dfrac{1}{3}\right) = 4$ and the standard deviation is $\sigma_X = \sqrt{np(1-p)} = \sqrt{(12)\left(\dfrac{1}{3}\right)\left(\dfrac{2}{3}\right)} \approx 1.6$.

   The distribution distribution will be clustered near the expected value of 4, and tail off on both sides. Choices A and C are too uniform and choice E is centered too far left. Choice D has too little variability. The simulation from choice B is the best choice. Alternately, you could calculate a few of the binomial probabilities $P(X = 0) = 0.008$, $P(X = 1) = 0.046$, $P(X = 2) = 0.127$. You could get all of the theoretical probabilities at once with the command binompdf$(12,1/3) \Rightarrow L_1$.

29. **(A)**

   Since the volunteers tried each formula, the design is matched pairs. Using paired data, the confidence interval will use $n - 1 = 10 - 1 = 9$ degrees of freedom.

30. **(C)**

   The population of interest is students that live in the dorms. Surveying only students in the cafeteria causes selection bias or undercoverage. Choices D and E are guilty of nonresponse or voluntary response bias. A random sample from the population of interest is what is needed.

31. (B)

The medians are clearly marked on boxplots. Choice A is wrong since we cannot know the total amount unless we knew individual data points. The skew on the boys' graph could put their mean above the girls', but we do not know for sure. The interquartile range is the width of the box—clearly more for the girls—so choice D is wrong. Choice E may be true, but we cannot know for sure unless we had the values of the data points.

32. (C)

The t-distribution is used for inference on means when the population standard deviation is not known.

33. (D)

The margin of error accounts for using sample data to estimate a population parameter. It has nothing to do with the population proportion itself, so choices A and B are wrong. We aim to be *within* a margin of error, not outside of it like choice C suggests. The margin of error is not a nonresponse rate, so choice E is wrong.

34. (B)

We are asked for an estimated *increase* in sales, not a total amount of sales. We simply need to multiply the slope by 3. Note that the unit of x is thousands of dollars.

35. (E)

A Type I error is committed when rejecting a true null hypothesis. Hypotheses for this situation are $H_0: \mu = 900$ and $H_a: \mu < 900$. So, a Type I error is concluding that μ is less than 900 when it really is not.

36. (A)

At first glance, this scenario may look like a systematic random sample, but it is unreasonable to think that the entire population of parents was present at the Open House. When you sample from the people at hand instead of the entire population of interest, you are doing a convenience sample.

37. **(A)**

   To meet the conditions for inference on slope, there must be no apparent pattern in the residuals plot, so choice B is incorrect. A histogram of residuals should look approximately unimodal and symmetric, with no outliers, so choices D and E are incorrect. Choice C is not good because the residuals plot does not have a consistent spread. The variability of the errors is not constant. Note that as the x-value increases, the spread of the residuals tends to decrease and then increases again.

38. **(E)**

   Random assignment must be present, so right away we can eliminate choices A, C, and D. Choice B is a nice design, but E is just better. A matched-pairs design allows for more powerful inference methods.

39. **(B)**

   If we let X = the gain/loss on one ticket, we can find the probability distribution of X. $P(X = -1) = 0.90$, since 9 out of 10 times the ticket will be a loser. $P(X = 23) = 0.01$, since the last two digits will both be "0" 1 out of 100 times. This leaves $P(X = 1) = 0.09$. Note that the possible values of X take into account the winnings and the \$1 for buying the ticket. We need to calculate the expected value of X. $E(X) = 0.90(-1) + 0.09(1) + 0.01(23) = -0.58$.

40. **(A)**

   This is the definition of the standard error of $\bar{x}$. Choice C is the sample standard deviation of x, not $\bar{x}$. Choice D is the population standard deviation of x.

Detailed Explanations of Answers

Section II

Free-Response Solutions

1.

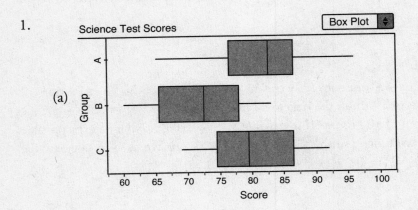

(a)

(b) The students from room A had a slightly higher median than the students in room C, 82.5 to 79.5. Room B's scores appear to be far lower than each of the other two, with a median 7 points lower than B, and more than half of its students scoring below the first quartile from each of the other rooms. All distributions are close to symmetric, with room A being slightly skewed left. The biggest difference among the rooms is the spread of scores. The scores in room A tended to spread out more than the scores in rooms B and C, with room A having a higher maximum and a lower minimum than room C.

2. (a) Let O = the event that a randomly selected American has O-type blood.

   Let N = the event that a randomly selected American has Rh-negative blood.

   If O and N are independent, then

   $$P(O \cap N) = P(O) \cdot P(N) = (0.45)(0.16) = 0.072$$

   (b) Given that $P(O \cap N) = 0.06$, we want to find $P(O \mid N)$.

   $$P(O \mid N) = \frac{P(O \cap N)}{P(N)} = \frac{0.06}{0.16} = 0.375.$$

(c) Let X = the number of people that donate until someone with O-negative does. X has a geometric probability distribution, with parameter $p = 0.06$. In a geometric probability distribution, $E(X) = \dfrac{1}{p} = \dfrac{1}{0.06} \approx 16.7$ people.

3. (a) Let X = age level and Y = mean score. The slope of a least-squares regression line is given by $b_1 = r\dfrac{s_y}{s_x} = (0.929)\left(\dfrac{7.75}{3.03}\right) \approx 2.376$. To find the y-intercept of the line, we use the fact that the point $(\bar{x}, \bar{y})$ is always on the least-squares regression line.

$$\hat{y} = b_0 + b_1 x$$
$$\bar{y} = b_0 + (2.376)\bar{x}$$
$$76 = b_0 + (2.376)(14.5)$$
$$a = 41.546.$$

So, the equation of the least-squares regression line is $\hat{y} = 41.546 + 2.376x$, or better yet.

$$\widehat{meanscore} = 41.546 + 2.376 \cdot age$$

(b) If $r = 0.929$, then $r^2 = 0.863$. This means that 86.3% of the variation in mean scores can be explained by the regression with age level.

(c) The residuals plot shows a curved pattern. The first three residuals are all negative, followed by four positive ones, and three more negative ones. Such a pattern suggests that a linear model is not best for these data.

4. (a) If μ_B = true mean weight of business passenger bags, and μ_P = true mean weight of pleasure passenger bags, then a 95% confidence interval for $\mu_B - \mu_P$ is

$$(\bar{x}_B - \bar{x}_P) \pm t^* \sqrt{\dfrac{s_B^2}{n_B} + \dfrac{s_P^2}{n_P}}$$

$$= (33.68 - 8.92) \pm 2.042\sqrt{\dfrac{4.62^2}{31} + \dfrac{7.42^2}{59}}$$

$$= 24.76 \pm 2.60$$

$$= (22.16, \ 27.32).$$

This calculation uses the t-critical value for 95% and 30 degrees of freedom, one less than the smaller sample size.

If done with a graphing calculator, the interval is (22.228, 27.292), and the calculator uses the t-critical value for 85 degrees of freedom.

We are 95% confident that business travelers' carry-on bags are on average between about 22.2 and 27.3 pounds heavier than pleasure travelers' carry-on bags.

(b) We can assume that we have two independent random samples since the passengers were selected randomly over a period of several weeks.

Each sample size is less than 10% of a potentially infinite population, so individual observations are independent of each other.

Although the dotplot for weights of bags of pleasure passengers is fairly right-skewed, having a sample size as large as 59 means that the sampling distribution for the sample mean of weights should be approximately normal. The dotplot for the weights of business passengers is skewed left somewhat, but again, the sample size is large enough that the sampling distribution of the sample mean will be approximately normal. Because of this, our conditions for inference, as described in part (a), have been met.

5. (a) The population of interest here is those students who own cars, not the entire student body. Therefore, the student body president should consider only those 900 students who own cars to be the sampling frame. Each of the 900 students should be assigned a three-digit number from 001 to 900; then three-digit numbers are taken from a random digit table (ignoring 901–999, 000, and repeats) until 100 students are chosen. This will yield an SRS of students who own cars.

    (b) The sample should be divided proportionally between sophomores, juniors, and seniors. Sophomores are $180/900 = 20\%$ of the students who own cars, juniors are 35%, and seniors are 45%. Therefore, a sample of 100 students should include 20 sophomores, 35 juniors, and 45 seniors. For the 180 sophomores with cars, assign each a number from 001 to 180 and then select 20 numbers at random without replacement. For the juniors, assign numbers 001–315 and select 35 numbers at random. For the seniors, assign the numbers 001–405 and select 45 numbers at random. Note how the stratified sample consists of three smaller simple random samples.

    (c) Both methods will give good and unbiased results. The president's simple random sample is a little easier to implement, requiring only one assignment of numbers. The secretary's stratified random sample requires doing this three times. The advantage of the secretary's plan is that stratification will prevent a disproportionate number of one grade from being sampled. The simple random sample does not guarantee proportional representation.

6. (a) In the series where the two teams are evenly matched:

    (1) Let the digits 0–4 represent a win for team A, and the digits 5–9 represent a win for team B.

    (2) Choose digits one at a time moving across the table, until a team has won four games.

    (3) Record the number of games the series lasted.

    (4) Repeat this simulation five times.

(b) In the series where one team wins 70% of the time:

   (1) Let the digits 0–6 will represent a win for team A, and 7–9 will represent a win for team B.

   Steps (2)–(4) will be identical to part (a).

(c) For evenly matched teams,

$$9301593 = BAAABBA = 7 \text{ games}$$
$$615034 = BABAAA = 6 \text{ games}$$
$$13722 = AABAA = 5 \text{ games}$$
$$9597 = BBBB = 4 \text{ games}$$
$$291363 = ABAABA = 6 \text{ games}$$

For team A winning 70% of games,

$$7714756 = BBAABAA = 7 \text{ games}$$
$$037685 = AABABA = 6 \text{ games}$$
$$38405 = ABAAA = 5 \text{ games}$$
$$5333 = AAAA = 4 \text{ games}$$
$$63186 = AAABA = 5 \text{ games}$$

(d) If teams are evenly matched, more games will need to be played to find a winner than when one team wins a majority of games played between them. Simulation 1 has generally longer series than simulation 2; therefore, simulation 1 is of teams that are evenly matched.

(e) The expected number of games for an evenly matched series as determined by simulation 1 is $(4)(0.12) + (5)(0.25) + (6)(0.33) + (7)(0.30) = 5.81$ games. The expected number of games for a series where one team wins 70% of the time as determined by simulation 2 is $(4)(0.25) + (5)(0.30) + (6)(0.27) + (7)(0.18) = 5.38$ games. Evenly matched series last, on average, 0.43 games longer.

Answer Sheet

Section I

1. Ⓐ Ⓑ Ⓒ Ⓓ Ⓔ
2. Ⓐ Ⓑ Ⓒ Ⓓ Ⓔ
3. Ⓐ Ⓑ Ⓒ Ⓓ Ⓔ
4. Ⓐ Ⓑ Ⓒ Ⓓ Ⓔ
5. Ⓐ Ⓑ Ⓒ Ⓓ Ⓔ
6. Ⓐ Ⓑ Ⓒ Ⓓ Ⓔ
7. Ⓐ Ⓑ Ⓒ Ⓓ Ⓔ
8. Ⓐ Ⓑ Ⓒ Ⓓ Ⓔ
9. Ⓐ Ⓑ Ⓒ Ⓓ Ⓔ
10. Ⓐ Ⓑ Ⓒ Ⓓ Ⓔ
11. Ⓐ Ⓑ Ⓒ Ⓓ Ⓔ
12. Ⓐ Ⓑ Ⓒ Ⓓ Ⓔ
13. Ⓐ Ⓑ Ⓒ Ⓓ Ⓔ
14. Ⓐ Ⓑ Ⓒ Ⓓ Ⓔ

15. Ⓐ Ⓑ Ⓒ Ⓓ Ⓔ
16. Ⓐ Ⓑ Ⓒ Ⓓ Ⓔ
17. Ⓐ Ⓑ Ⓒ Ⓓ Ⓔ
18. Ⓐ Ⓑ Ⓒ Ⓓ Ⓔ
19. Ⓐ Ⓑ Ⓒ Ⓓ Ⓔ
20. Ⓐ Ⓑ Ⓒ Ⓓ Ⓔ
21. Ⓐ Ⓑ Ⓒ Ⓓ Ⓔ
22. Ⓐ Ⓑ Ⓒ Ⓓ Ⓔ
23. Ⓐ Ⓑ Ⓒ Ⓓ Ⓔ
24. Ⓐ Ⓑ Ⓒ Ⓓ Ⓔ
25. Ⓐ Ⓑ Ⓒ Ⓓ Ⓔ
26. Ⓐ Ⓑ Ⓒ Ⓓ Ⓔ
27. Ⓐ Ⓑ Ⓒ Ⓓ Ⓔ
28. Ⓐ Ⓑ Ⓒ Ⓓ Ⓔ

29. Ⓐ Ⓑ Ⓒ Ⓓ Ⓔ
30. Ⓐ Ⓑ Ⓒ Ⓓ Ⓔ
31. Ⓐ Ⓑ Ⓒ Ⓓ Ⓔ
32. Ⓐ Ⓑ Ⓒ Ⓓ Ⓔ
33. Ⓐ Ⓑ Ⓒ Ⓓ Ⓔ
34. Ⓐ Ⓑ Ⓒ Ⓓ Ⓔ
35. Ⓐ Ⓑ Ⓒ Ⓓ Ⓔ
36. Ⓐ Ⓑ Ⓒ Ⓓ Ⓔ
37. Ⓐ Ⓑ Ⓒ Ⓓ Ⓔ
38. Ⓐ Ⓑ Ⓒ Ⓓ Ⓔ
39. Ⓐ Ⓑ Ⓒ Ⓓ Ⓔ
40. Ⓐ Ⓑ Ⓒ Ⓓ Ⓔ

FREE-RESPONSE ANSWER SHEET

For the free-response section, write your answers on sheets of blank paper.

A grid for Question 1 and the random digit table for Question 6 are provided below.

93015	93615	03413	72295	97291
36377	14756	03768	53840	55333
63186	77781	69103	43259	01660
23825	65704	49525	84121	44856
97187	05901	61053	04173	07717

Appendix A
Formulas and Tables

I. Descriptive Statistics

$$\bar{x} = \frac{\sum x_i}{n}$$

$$s_x = \sqrt{\frac{1}{n-1}\sum(x_i - \bar{x})^2}$$

$$s_p = \sqrt{\frac{(n_1-1)s_1^2 + (n_2-1)s_2^2}{(n_1-1)+(n_2-1)}}$$

$$\hat{y} = b_0 + b_1 x$$

$$b_1 = \frac{\sum(x_i - \bar{x})(y_i - \bar{y})}{\sum(x_i - \bar{x})^2}$$

$$b_0 = \bar{y} - b_1\bar{x}$$

$$r = \frac{1}{n-1}\sum\left(\frac{x_i - \bar{x}}{s_x}\right)\left(\frac{y_i - \bar{y}}{s_y}\right)$$

$$b_1 = r\frac{s_y}{s_x}$$

$$s_{b_1} = \frac{\sqrt{\dfrac{\sum(y_i - \hat{y}_i)^2}{n-2}}}{\sqrt{\sum(x_i - \bar{x})^2}}.$$

II. Probability

$$P(A \cup B) = P(A) + P(B) - P(A \cap B)$$

$$P(A \mid B) = \frac{P(A \cap B)}{P(B)}$$

$$E(X) = \mu_x = \sum x_i p_i$$

$$\mathrm{Var}(X) = \sigma_x^2 = \sum (x_i - \mu_x)^2 \, p_i.$$

If X has a binomial distribution with parameters n and p, then

$$P(X = k) = \binom{n}{k} p^k (1 - p)^{n-k}$$

$$\mu_x = np$$

$$\sigma_x = \sqrt{np(1 - p)}$$

$$\mu_{\hat{p}} = p$$

$$\sigma_{\hat{p}} = \sqrt{\frac{p(1 - p)}{n}}.$$

If $\bar{x}$ is the mean of a random sample of size n from an infinite population with mean μ and standard deviation σ, then

$$\mu_{\bar{x}} = \mu$$

$$\sigma_{\bar{x}} = \frac{\sigma}{\sqrt{n}}.$$

III. Inferential Statistics

$$\text{Standardized test statistic} = \frac{\text{statistic} - \text{parameter}}{\text{standard deviation of statistic}}$$

$$\text{Confidence interval} = \text{statistic} \pm (\text{critical value}) \cdot (\text{standard deviation of statistic}).$$

Single-Sample

Statistic	Standard Deviation of Statistic
Sample Mean	$\dfrac{\sigma}{\sqrt{n}}$
Sample Proportion	$\sqrt{\dfrac{p(1-p)}{n}}$

Two-Sample

Statistic	Standard Deviation of Statistic
Difference of sample means	$\sqrt{\dfrac{\sigma_1^2}{n_1} + \dfrac{\sigma_2^2}{n_2}}$ Special case when $\sigma_1 = \sigma_2$ $\sigma\sqrt{\dfrac{1}{n_1} + \dfrac{1}{n_2}}$
Difference of sample proportions	$\sqrt{\dfrac{p_1(1-p_1)}{n_1} + \dfrac{p_2(1-p_2)}{n_2}}$ Special case when $p_1 = p_2$ $\sqrt{p(1-p)}\sqrt{\dfrac{1}{n_1} + \dfrac{1}{n_2}}$

$$\text{Chi-square test statistic} = \sum \frac{(\text{observed} - \text{expected})^2}{\text{expected}}$$

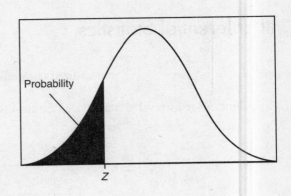

Table entry for z is the probability lying below z.

Table A						Standard normal probabilities				
Z	0.00	0.01	0.02	0.03	0.04	0.05	0.06	0.07	0.08	0.09
−3.4	0.0003	0.0003	0.0003	0.0003	0.0003	0.0003	0.0003	0.0003	0.0003	0.0002
−3.3	0.0005	0.0005	0.0005	0.0004	0.0004	0.0004	0.0004	0.0004	0.0004	0.0003
−3.2	0.0007	0.0007	0.0006	0.0006	0.0006	0.0006	0.0006	0.0005	0.0005	0.0005
−3.1	0.0010	0.0009	0.0009	0.0009	0.0008	0.0008	0.0008	0.0008	0.0007	0.0007
−3.0	0.0013	0.0013	0.0013	0.0012	0.0012	0.0011	0.0011	0.0011	0.0010	0.0010
−2.9	0.0019	0.0018	0.0018	0.0017	0.0016	0.0016	0.0015	0.0015	0.0014	0.0014
−2.8	0.0026	0.0025	0.0024	0.0023	0.0023	0.0022	0.0021	0.0021	0.0020	0.0019
−2.7	0.0035	0.0034	0.0033	0.0032	0.0031	0.0030	0.0029	0.0028	0.0027	0.0026
−2.6	0.0047	0.0045	0.0044	0.0043	0.0041	0.0040	0.0039	0.0038	0.0037	0.0036
−2.5	0.0062	0.0060	0.0059	0.0057	0.0055	0.0054	0.0052	0.0051	0.0049	0.0048
−2.4	0.0082	0.0080	0.0078	0.0075	0.0073	0.0071	0.0069	0.0068	0.0066	0.0064
−2.3	0.0107	0.0104	0.0102	0.0099	0.0096	0.0094	0.0091	0.0089	0.0087	0.0084
−2.2	0.0139	0.0136	0.0132	0.0129	0.0125	0.0122	0.0119	0.0116	0.0113	0.0110
−2.1	0.0179	0.0174	0.0170	0.0166	0.0162	0.0158	0.0154	0.0150	0.0146	0.0143
−2.0	0.0228	0.0222	0.0217	0.0212	0.0207	0.0202	0.0197	0.0192	0.0188	0.0183
−1.9	0.0287	0.0281	0.0274	0.0268	0.0262	0.0256	0.0250	0.0244	0.0239	0.0233
−1.8	0.0359	0.0351	0.0344	0.0336	0.0329	0.0322	0.0314	0.0307	0.0301	0.0294
−1.7	0.0446	0.0436	0.0427	0.0418	0.0409	0.0401	0.0392	0.0384	0.0375	0.0367
−1.6	0.0548	0.0537	0.0526	0.0516	0.0505	0.0495	0.0485	0.0475	0.0465	0.0455
−1.5	0.0668	0.0655	0.0643	0.0630	0.0618	0.0606	0.0594	0.0582	0.0571	0.0559
−1.4	0.0808	0.0793	0.0778	0.0764	0.0749	0.0735	0.0721	0.0708	0.0694	0.0681
−1.3	0.0968	0.0951	0.0934	0.0918	0.0901	0.0885	0.0869	0.0853	0.0838	0.0823
−1.2	0.1151	0.1131	0.1112	0.1093	0.1075	0.1056	0.1038	0.1020	0.1003	0.0985
−1.1	0.1357	0.1335	0.1314	0.1292	0.1271	0.1251	0.1230	0.1210	0.1190	0.1170
−1.0	0.1587	0.1562	0.1539	0.1515	0.1492	0.1469	0.1446	0.1423	0.1401	0.1379
−0.9	0.1841	0.1814	0.1788	0.1762	0.1736	0.1711	0.1685	0.1660	0.1635	0.1611

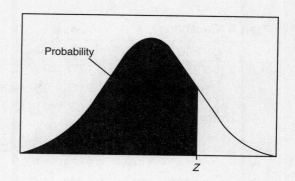

Probability

Table entry for *z* is the probability lying below *z*.

Table A (*Continued*)

Z	0.00	0.01	0.02	0.03	0.04	0.05	0.06	0.07	0.08	0.09
−0.8	0.2119	0.2090	0.2061	0.2033	0.2005	0.1977	0.1949	0.1922	0.1894	0.1867
−0.7	0.2420	0.2389	0.2358	0.2327	0.2296	0.2266	0.2236	0.2206	0.2177	0.2148
−0.6	0.2743	0.2709	0.2676	0.2643	0.2611	0.2578	0.2546	0.2514	0.2483	0.2451
−0.5	0.3085	0.3050	0.3015	0.2981	0.2946	0.2912	0.2877	0.2843	0.2810	0.2776
−0.4	0.3446	0.3409	0.3372	0.3336	0.3300	0.3264	0.3228	0.3192	0.3156	0.3121
−0.3	0.3821	0.3783	0.3745	0.3707	0.3669	0.3632	0.3594	0.3557	0.3520	0.3483
−0.2	0.4207	0.4168	0.4129	0.4090	0.4052	0.4013	0.3974	0.3936	0.3897	0.3859
−0.1	0.4602	0.4562	0.4522	0.4483	0.4443	0.4404	0.4364	0.4325	0.4286	0.4247
−0.0	0.5000	0.4960	0.4920	0.4880	0.4840	0.4801	0.4761	0.4721	0.4681	0.4641
0.0	0.5000	0.5040	0.5080	0.5120	0.5160	0.5199	0.5239	0.5279	0.5319	0.5359
0.1	0.5398	0.5438	0.5478	0.5517	0.5557	0.5596	0.5636	0.5675	0.5714	0.5753
0.2	0.5793	0.5832	0.5871	0.5910	0.5948	0.5987	0.6026	0.6064	0.6103	0.6141
0.3	0.6179	0.6217	0.6255	0.6293	0.6331	0.6368	0.6406	0.6443	0.6480	0.6517
0.4	0.6554	0.6591	0.6628	0.6664	0.6700	0.6736	0.6772	0.6808	0.6844	0.6879
0.5	0.6915	0.6950	0.6985	0.7019	0.7054	0.7088	0.7123	0.7157	0.7190	0.7224
0.6	0.7257	0.7291	0.7324	0.7357	0.7389	0.7422	0.7454	0.7486	0.7517	0.7549
0.7	0.7580	0.7611	0.7642	0.7673	0.7704	0.7734	0.7764	0.7794	0.7823	0.7852
0.8	0.7881	0.7910	0.7939	0.7967	0.7995	0.8023	0.8051	0.8078	0.8106	0.8133
0.9	0.8159	0.8186	0.8212	0.8238	0.8264	0.8289	0.8315	0.8340	0.8365	0.8389
1.0	0.8413	0.8438	0.8461	0.8485	0.8508	0.8531	0.8554	0.8577	0.8599	0.8621
1.1	0.8643	0.8665	0.8686	0.8708	0.8729	0.8749	0.8770	0.8790	0.8810	0.8830
1.2	0.8849	0.8869	0.8888	0.8907	0.8925	0.8944	0.8962	0.8980	0.8997	0.9015
1.3	0.9032	0.9049	0.9066	0.9082	0.9099	0.9115	0.9131	0.9147	0.9162	0.9177
1.4	0.9192	0.9207	0.9222	0.9236	0.9251	0.9265	0.9279	0.9292	0.9306	0.9319
1.5	0.9332	0.9345	0.9357	0.9370	0.9382	0.9394	0.9406	0.9418	0.9429	0.9441
1.6	0.9452	0.9463	0.9474	0.9484	0.9495	0.9505	0.9515	0.9525	0.9535	0.9545

Table A (*Continued*)

Z	0.00	0.01	0.02	0.03	0.04	0.05	0.06	0.07	0.08	0.09
1.7	0.9554	0.9564	0.9573	0.9582	0.9591	0.9599	0.9608	0.9616	0.9625	0.9633
1.8	0.9641	0.9649	0.9656	0.9664	0.9671	0.9678	0.9686	0.9693	0.9699	0.9706
1.9	0.9713	0.9719	0.9726	0.9732	0.9738	0.9744	0.9750	0.9756	0.9761	0.9767
2.0	0.9772	0.9778	0.9783	0.9788	0.9793	0.9798	0.9803	0.9808	0.9812	0.9817
2.1	0.9821	0.9826	0.9830	0.9834	0.9838	0.9842	0.9846	0.9850	0.9854	0.9857
2.2	0.9861	0.9864	0.9868	0.9871	0.9875	0.9878	0.9881	0.9884	0.9887	0.9890
2.3	0.9893	0.9896	0.9898	0.9901	0.9904	0.9906	0.9909	0.9911	0.9913	0.9916
2.4	0.9918	0.9920	0.9922	0.9925	0.9927	0.9929	0.9931	0.9932	0.9934	0.9936
2.5	0.9938	0.9940	0.9941	0.9943	0.9945	0.9946	0.9948	0.9949	0.9951	0.9952
2.6	0.9953	0.9955	0.9956	0.9957	0.9959	0.9960	0.9961	0.9962	0.9963	0.9964
2.7	0.9965	0.9966	0.9967	0.9968	0.9969	0.9970	0.9971	0.9972	0.9973	0.9974
2.8	0.9974	0.9975	0.9976	0.9977	0.9977	0.9978	0.9979	0.9979	0.9980	0.9981
2.9	0.9981	0.9982	0.9982	0.9983	0.9984	0.9984	0.9985	0.9985	0.9986	0.9986
3.0	0.9987	0.9987	0.9987	0.9988	0.9988	0.9989	0.9989	0.9989	0.9990	0.9990
3.1	0.9990	0.9991	0.9991	0.9991	0.9992	0.9992	0.9992	0.9992	0.9993	0.9993
3.2	0.9993	0.9993	0.9994	0.9994	0.9994	0.9994	0.9994	0.9995	0.9995	0.9995
3.3	0.9995	0.9995	0.9995	0.9996	0.9996	0.9996	0.9996	0.9996	0.9996	0.9997
3.4	0.9997	0.9997	0.9997	0.9997	0.9997	0.9997	0.9997	0.9997	0.9997	0.9998

Probability p

Table entry for p and C is the point t^* with probability p lying above it and probability C lying between $-t^*$ and t^*.

Table B   **t-distribution critical values**

df	0.25	0.20	0.15	0.10	0.05	0.025	0.02	0.01	0.005	0.0025	0.001	0.0005
1	1.000	1.376	1.963	3.078	6.314	12.71	15.89	31.82	63.66	127.3	318.3	636.6
2	0.816	1.061	1.386	1.886	2.920	4.303	4.849	6.965	9.925	14.09	22.33	31.60
3	0.765	0.978	1.250	1.638	2.353	3.182	3.482	4.541	5.841	7.453	10.21	12.92
4	0.741	0.941	1.190	1.533	2.132	2.776	2.999	3.747	4.604	5.598	7.173	8.610
5	0.727	0.920	1.156	1.476	2.015	2.571	2.757	3.365	4.032	4.773	5.893	6.869

Tail probability p

Table B *Continued*

df	0.25	0.20	0.15	0.10	0.05	0.025	0.02	0.01	0.005	0.0025	0.001	0.0005
6	0.718	0.906	1.134	1.440	1.943	2.447	2.612	3.143	3.707	4.317	5.208	5.959
7	0.711	0.896	1.119	1.415	1.895	2.365	2.517	2.998	3.499	4.029	4.785	5.408
8	0.706	0.889	1.108	1.397	1.860	2.306	2.449	2.896	3.355	3.833	4.501	5.041
9	0.703	0.883	1.100	1.383	1.833	2.262	2.398	2.821	3.250	3.690	4.297	4.781
10	0.700	0.879	1.093	1.372	1.812	2.228	2.359	2.764	3.169	3.581	4.144	4.587
11	0.697	0.876	1.088	1.363	1.796	2.201	2.328	2.718	3.106	3.497	4.025	4.437
12	0.695	0.873	1.083	1.356	1.782	2.179	2.303	2.681	3.055	3.428	3.930	4.318
13	0.694	0.870	1.079	1.350	1.771	2.160	2.282	2.650	3.012	3.372	3.852	4.221
14	0.692	0.868	1.076	1.345	1.761	2.145	2.264	2.624	2.977	3.326	3.787	4.140
15	0.691	0.866	1.074	1.341	1.753	2.131	2.249	2.602	2.947	3.286	3.733	4.073
16	0.690	0.865	1.071	1.337	1.746	2.120	2.235	2.583	2.921	3.252	3.686	4.015
17	0.689	0.863	1.069	1.333	1.740	2.110	2.224	2.567	2.898	3.222	3.646	3.965
18	0.688	0.862	1.067	1.330	1.734	2.101	2.214	2.552	2.878	3.197	3.611	3.922
19	0.688	0.861	1.066	1.328	1.729	2.093	2.205	2.539	2.861	3.174	3.579	3.883
20	0.687	0.860	1.064	1.325	1.725	2.086	2.197	2.528	2.845	3.153	3.552	3.850
21	0.686	0.859	1.063	1.323	1.721	2.080	2.189	2.518	2.831	3.135	3.527	3.819
22	0.686	0.858	1.061	1.321	1.717	2.074	2.183	2.508	2.819	3.119	3.505	3.792
23	0.685	0.858	1.060	1.319	1.714	2.069	2.177	2.500	2.807	3.104	3.485	3.768
24	0.685	0.857	1.059	1.318	1.711	2.064	2.172	2.492	2.797	3.091	3.467	3.745
25	0.684	0.856	1.058	1.316	1.708	2.060	2.167	2.485	2.787	3.078	3.450	3.725
26	0.684	0.856	1.058	1.315	1.706	2.056	2.162	2.479	2.779	3.067	3.435	3.707
27	0.684	0.855	1.057	1.314	1.703	2.052	2.158	2.473	2.771	3.057	3.421	3.690
28	0.683	0.855	1.056	1.313	1.701	2.048	2.154	2.467	2.763	3.047	3.408	3.674
29	0.683	0.854	1.055	1.311	1.699	2.045	2.150	2.462	2.756	3.038	3.396	3.659
30	0.683	0.854	1.055	1.310	1.697	2.042	2.147	2.457	2.750	3.030	3.385	3.646
40	0.681	0.851	1.050	1.303	1.684	2.021	2.123	2.423	2.704	2.971	3.307	3.551
50	0.679	0.849	1.047	1.299	1.676	2.009	2.109	2.403	2.678	2.937	3.261	3.496
60	0.679	0.848	1.045	1.296	1.671	2.000	2.099	2.390	2.660	2.915	3.232	3.460
80	0.678	0.846	1.043	1.292	1.664	1.990	2.088	2.374	2.639	2.887	3.195	3.416
100	0.677	0.845	1.042	1.290	1.660	1.984	2.081	2.364	2.626	2.871	3.174	3.390
1000	0.675	0.842	1.037	1.282	1.646	1.962	2.056	2.330	2.581	2.813	3.098	3.300
∞	0.674	0.841	1.036	1.282	1.645	1.960	2.054	2.326	2.576	2.807	3.091	3.291
	50%	60%	70%	80%	90%	95%	96%	98%	99%	99.5%	99.8%	99.9%

Confidence level C

Table entry for p is the point (χ^2) with probability p lying above it

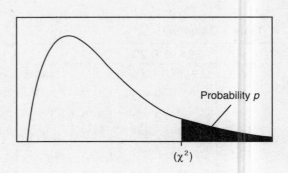

Probability p

(χ^2)

Table C χ^2-**critical values**

	Tail probability p										
df	0.25	0.20	0.15	0.10	0.05	0.025	0.02	0.01	0.005	0.0025	0.001
1	1.32	1.64	2.07	2.71	3.84	5.02	5.41	6.63	7.88	9.14	10.83
2	2.77	3.22	3.79	4.61	5.99	7.38	7.82	9.21	10.60	11.98	13.82
3	4.11	4.64	5.32	6.25	7.81	9.35	9.84	11.34	12.84	14.32	16.27
4	5.39	5.99	6.74	7.78	9.49	11.14	11.67	13.28	14.86	16.42	18.47
5	6.63	7.29	8.12	9.24	11.07	12.83	13.39	15.09	16.75	18.39	20.51
6	7.84	8.56	9.45	10.64	12.59	14.45	15.03	16.81	18.55	20.25	22.46
7	9.04	9.80	10.75	12.02	14.07	16.01	16.62	18.48	20.28	22.04	24.32
8	10.22	11.03	12.03	13.36	15.51	17.53	18.17	20.09	21.95	23.77	26.12
9	11.39	12.24	13.29	14.68	16.92	19.02	19.68	21.67	23.59	25.46	27.88
10	12.55	13.44	14.53	15.99	18.31	20.48	21.16	23.21	25.19	27.11	29.59
11	13.70	14.63	15.77	17.28	19.68	21.92	22.62	24.72	26.76	28.73	31.26
12	14.85	15.81	16.99	18.55	21.03	23.34	24.05	26.22	28.30	30.32	32.91
13	15.98	16.98	18.20	19.81	22.36	24.74	25.47	27.69	29.82	31.88	34.53
14	17.12	18.15	19.41	21.06	23.68	26.12	26.87	29.14	31.32	33.43	36.12
15	18.25	19.31	20.60	22.31	25.00	27.49	28.26	30.58	32.80	34.95	37.70
16	19.37	20.47	21.79	23.54	26.30	28.85	29.63	32.00	34.27	36.46	39.25
17	20.49	21.61	22.98	24.77	27.59	30.19	31.00	33.41	35.72	37.95	40.79
18	21.60	22.76	24.16	25.99	28.87	31.53	32.35	34.81	37.16	39.42	42.31
19	22.72	23.90	25.33	27.20	30.14	32.85	33.69	36.19	38.58	40.88	43.82
20	23.83	25.04	26.50	28.41	31.41	34.17	35.02	37.57	40.00	42.34	45.31
21	24.93	26.17	27.66	29.62	32.67	35.48	36.34	38.93	41.40	43.78	46.80
22	26.04	27.30	28.82	30.81	33.92	36.78	37.66	40.29	42.80	45.20	48.27
23	27.14	28.43	29.98	32.01	35.17	38.08	38.97	41.64	44.18	46.62	49.73
24	28.24	29.55	31.13	33.20	36.42	39.36	40.27	42.98	45.56	48.03	51.18
25	29.34	30.68	32.28	34.38	37.65	40.65	41.57	44.31	46.93	49.44	52.62

Table C (*Continued*)

df	0.25	0.20	0.15	0.10	0.05	0.025	0.02	0.01	0.005	0.0025	0.001
26	30.43	31.79	33.43	35.56	38.89	41.92	42.86	45.64	48.29	50.83	54.05
27	31.53	32.91	34.57	36.74	40.11	43.19	44.14	46.96	49.64	52.22	55.48
28	32.62	34.03	35.71	37.92	41.34	44.46	45.42	48.28	50.99	53.59	56.89
29	33.71	35.14	36.85	39.09	42.56	45.72	46.69	49.59	52.34	54.97	58.30
30	34.80	36.25	37.99	40.26	43.77	46.98	47.96	50.89	53.67	56.33	59.70
40	45.62	47.27	49.24	51.81	55.76	59.34	60.44	63.69	66.77	69.70	73.40
50	56.33	58.16	60.35	63.17	67.50	71.42	72.61	76.15	79.49	82.66	86.66
60	66.98	68.97	71.34	74.40	79.08	83.30	84.58	88.38	91.95	95.34	99.61
80	88.13	90.41	93.11	96.58	101.9	106.6	108.1	112.3	116.3	120.1	124.8
100	109.1	111.7	114.7	118.5	124.3	129.6	131.1	135.8	140.2	144.3	149.4

Glossary

Addition rule $P(A \cap B) = P(A) + P(B) - P(A \cup B)$ aids in computing the chances of one of several events occurring at a given time.

Alpha (α) the probability of a Type I error. See significance level.

Alternative hypothesis the hypothesis stating what the researcher is seeking evidence of. A statement of inequality. It can be written looking for the difference or change in one direction from the null hypothesis or both.

Association relationship between or among variables.

Back-transform the process by which values are substituted into a model of transformed data, and then reversing the transforming process to obtain the predicted value or model for nontransformed data.

Bar chart a graphical display used with categorical data, where frequencies for each category are shown in vertical bars.

Bell-shaped often used to describe the normal distribution. See mound-shaped.

Beta (β) The probability of a Type II error. See power.

Bias the term for systematic deviation from the truth (parameter), caused by systematically favoring some outcomes over others.

Biased a sampling method is biased if it tends to produce samples that do not represent the population.

Bimodal a distribution with two clear peaks.

Binomial distribution the probability distribution of a binomial random variable.

Binomial random variable a random variable X (a) that has a fixed number of trials of a random phenomenon n, (b) that has only two possible outcomes on each trial, (c) for which the probability of a success is constant for each trial, and (d) for which each trial is independent of other trials.

Bins the intervals that define the "bars" of a histogram.

Bivariate data consists of two variables, an explanatory and a response variable, usually quantitative.

Blinding practice of denying knowledge to subjects about which treatment is imposed upon them.

Blocks subgroups of the experimental units that are separated by some characteristic before treatments are assigned because they may respond differently to the treatments.

Box-and-whisker plot/boxplot a graphical display of the five-number summary of a set of data, which also shows outliers.

Categorical variable a variable recorded as labels, names, or other non-numerical outcomes.

Census a study that observes, or attempts to observe, every individual in a population.

Central limit theorem as the size n of a simple random sample increases, the shape of the sampling distribution of tends toward being normally distributed.

Chance device a mechanism used to determine random outcomes.

Cluster sample a sample in which a simple random sample of heterogeneous subgroups of a population is selected.

Clusters heterogeneous subgroups of a population.

Coefficient of determination (r^2) percent of variation in the response variable explained by its linear relationship with the explanatory variable.

Complement of an event is the set of all outcomes of an experiment that are not included in the event.

Complementary events two events whose probabilities add up to 1.

Completely randomized design one in which all experimental units are assigned treatments solely by chance.

Conditional distribution see conditional frequencies.

Conditional frequencies relative frequencies for each cell in a two-way table relative to one variable.

Conditional probability the probability of an event occurring given that another has occurred. The probability of A given that B has occurred is denoted as $P(A|B)$.

Confidence intervals give an estimated range that is likely to contain an unknown population parameter.

Confidence level the level of certainty that a population parameter exists in the calculated confidence interval.

Confounding the situation where the effects of two or more explanatory variables on the response variable cannot be separated.

Confounding variable a variable whose effect on the response variable cannot be untangled from the effects of the treatment.

Contingency table see two-way table.

Continuous random variables those typically found by measuring, such as heights or temperatures.

Control group a baseline group that may be given no treatment, a faux treatment like a placebo, or an accepted treatment that is to be compared to another.

Control the principle that potential sources of variation due to variables not under consideration must be reduced.

Convenience sample composed of individuals who are easily accessed or contacted.

Correlation coefficient (r) a measure of the strength of a linear relationship,

$$r = \frac{1}{n-1} \sum \left(\frac{x_i - \bar{x}}{s_x} \right) \left(\frac{y_i - \bar{y}}{s_y} \right).$$

Critical value the value that the test statistic must exceed in order to reject the null hypothesis. When computing a confidence interval, the value of t^* (or z^*) where $\pm\, t^*$ (or $\pm\, z^*$) bounds the central $C\%$ of the t (or z) distribution.

Cumulative frequency the sums of the frequencies of the data values from smallest to largest.

Data set collection of observations from a sample or population.

Dependent events two events are called dependent when they are related and the fact that one event has occurred changes the probability that the second event occurs.

Discrete random variables those usually obtained by counting.

Disjoint events events that cannot occur simultaneously.

Distribution frequencies of values in a data set.

Dotplot a graphical display used with univariate data. Each data point is shown as a dot located above its numerical value on the horizontal axis.

Double-blind when both the subjects and data gatherers are ignorant about which treatment a subject received.

Empirical rule (68-95-99.7 rule) gives benchmarks for understanding how probability is distributed under a normal curve. In the normal distribution,

68% of the observations are within one standard deviation of the mean, 95% is within two standard deviations of the mean, and 99.7% is within three standard deviations of the mean.

Estimation the process of determining the value of a population parameter from a sample statistic.

Expected value the mean of a probability distribution.

Experiment a study where the researcher deliberately influences individuals by imposing conditions and determining the individuals' responses to those conditions.

Experimental units individuals in an experiment.

Explanatory variable explains the response variable, sometimes known as the treatment variable.

Exponential model a model of the form of $y = ab^x$.

Extrapolation using a model to predict values far outside the range of the explanatory variable, which is prone to creating unreasonable predictions.

Factors one or more explanatory variables in an experiment.

First quartile symbolized Q_1, represents the median of the lower 50% of a data set.

Five-number summary the minimum, first quartile (Q_1), median, third quartile (Q_3), and maximum values in a data set.

Frequency table a display organizing categorical or numerical data and how often each occurs.

Geometric distribution the probability distribution of a geometric random variable X. All possible outcomes of X before the first success is seen and their associated probabilities.

Geometric random variable a random variable X (a) that has two possible outcomes of each trial, (b) for which the probability of a success is

constant for each trial, and (c) for which each trial is independent of the other trials.

Graphical display a visual representation of a distribution.

Histogram used with univariate data, frequencies are shown on the vertical axis, and intervals or bins define the values on the horizontal axis.

Independent events two events are called independent when knowing that one event has occurred does not change the probability that the second event occurs.

Independent random variables if the values of one random variable have no association with the values of another, the two variables are called independent random variables.

Influential point an extreme value whose removal would drastically change the slope of the least-squares regression model.

Interquartile range describes the spread of middle 50% of a data set, $IQR = Q_3 - Q_1$.

Joint distribution see joint frequencies.

Joint frequencies frequencies for each cell in a two-way table relative to the total number of data.

Law of large numbers the long-term relative frequency of an event gets closer to the true relative frequency as the number of trials of a random phenomenon increases.

Least-squares regression line (LSRL) the "best-fit" line that is calculated by minimizing the sum of the squares of the differences between the observed and predicted values of the line. The LSRL has the equation $\hat{y} = b_0 + b_1 x$.

Levels the different quantities or categories of a factor in an experiment.

Linear regression a method of finding the best model for a linear relationship between the explanatory and response variable.

Logarithmic transformation procedure that changes a variable by taking the logarithm of each of its values.

Lurking variable a variable that has an effect on the outcome of a study but was not part of the investigation.

Margin of error a range of values to the left and right of a point estimate.

Marginal distribution see marginal frequencies.

Marginal frequencies row totals and column totals in a two-way table.

Matched-pairs design the design of a study where experimental units are naturally paired by a common characteristic, or with themselves in a before–after type of study.

Maximum the largest numerical value in a data set.

Mean the arithmetic average of a data set; the sum of all the values divided by the number of values, $\bar{x} = \dfrac{\sum x_i}{n}$.

Mean of a binomial random variable X $\mu_x = np$.

Mean$_n$ of a discrete random variable $\mu_x = \sum_{i=1}^{n} x_i P(x_i)$.

Mean of a geometric random variable $\mu_x = \dfrac{1}{p}$.

Measures of center these locate the middle of a distribution. The mean and median are measures of center.

Median the middle value of a data set; the equal areas point, where 50% of the data are at or below this value, and 50% of the data are at or above this value.

Minimum the smallest numerical value in a data set.

Mound-shaped resembles a hill or mound; a distribution that is symmetric and unimodal.

Multiplication rule $P(A \cup B) = P(A) \cdot P(B|A)$ is used when we are interested in the probability of two events occurring simultaneously, or in succession.

Multistage sample a sample resulting from multiple applications of cluster, stratified, and/or simple random sampling.

Mutually exclusive events see disjoint events.

Nonresponse bias the situation where an individual selected to be in the sample is unwilling, or unable, to provide data.

Normal distribution a continuous probability distribution that appears in many situations, both natural and man-made. It has a bell-shape and the area under the normal density curve is always equal to 1.

Null hypothesis the hypothesis of no difference, no change, and no association. A statement of equality, usually written in the form H_0: *parameter* = hypothesized value.

Observational study attempts to determine relationships between variables, but the researcher imposes no conditions as in an experiment.

Observed values actual outcomes or data from a study or an experiment.

One-way table a frequency table of one variable.

Outlier an extreme value in a data set. Quantified by being less than $Q_1 - 1.5IQR$ or more than $Q_3 + 1.5IQR$.

Percentiles divide the data set into 100 equal parts. An observation at the Pth percentile is higher than P percent of all observations.

Placebo a faux treatment given in an experiment that resembles the real treatment under consideration.

Placebo effect a phenomenon where subjects show a response to a treatment merely because the treatment is imposed regardless of its actual effect.

Point estimate an approximate value that has been calculated for the unknown parameter.

Population the collection of all individuals under consideration in a study.

Population parameter a characteristic or measure of a population.

Position location of a data value relative to the population.

Power the probability of correctly rejecting the null hypothesis when it is in fact false. Equal to $1 - \beta$. See beta and Type II error.

Power model a function in the form of $y = ax^b$.

Predicted value the value of the response variable predicted by a model for a given explanatory variable.

Probability describes the chance that a certain outcome of a random phenomenon will occur.

Probability distribution a discrete random variable X is a function of all n possible outcomes of the random variable (x_i) and their associated probabilities $P(x_i)$.

Probability sample composed of individuals selected by chance.

P-value the probability of observing a test statistic as extreme as, or more extreme than, the statistic obtained from a sample, under the assumption that the null hypothesis is true.

Quantitative a variable whose values are counts or measurements.

Random digit table a chance device that is used to select experimental units or conduct simulations.

Random phenomena those outcomes that are unpredictable in the short term, but nevertheless, have a long-term pattern.

Random sample a sample composed of individuals selected by chance.

Random variables numerical outcome of a random phenomenon.

Randomization the process by which treatments are assigned by a chance mechanism to the experimental units.

Randomized block design first, units are sorted into subgroups or blocks, and then treatments are randomly assigned within the blocks.

Range calculated as the maximum value minus the minimum value in a data set.

Relative frequency percentage or proportion of the whole number of data.

Relative frequency segmented bar chart a method of graphing a conditional distribution.

Replication the practice of reducing chance variation by assigning each treatment to many experimental units.

Residual observed value minus predicted value of the response variable.

Response bias because of the manner in which an interview is conducted, because of the phrasing of questions, or because of the attitude of the respondent, inaccurate data are collected.

Response variable measures the outcomes that have been observed.

Sample a selected subset of a population from which data are gathered.

Sample statistic result of a sample used to estimate a parameter.

Sample survey a study that collects information from a sample of a population in order to determine one or more characteristics of the population.

Sampling distribution the probability distribution of a sample statistic when a sample is drawn from a population.

Sampling distribution of the sample mean $\overline{x}$ the distribution of sample means from all possible simple random samples of size n taken from a population.

Sampling distribution of a sample proportion $\hat{p}$ the distribution of sample proportions from all possible simple random samples of size n taken from a population.

Sampling error see sampling variability.

Sampling variability natural variability due to the sampling process. Each possible random sample from a population will generate a different sample statistic.

Scatterplots used to visualize bivariate data. The explanatory variable is shown on the horizontal axis and the response variable is shown on the vertical axis.

Significance level the probability of a Type I error. A benchmark against which the P-value compared to determine if the null hypothesis will be rejected. See also alpha. (α).

Simple random sample (SRS) a sample where n individuals are selected from a population in a way that every possible combination of n individuals is equally likely.

Simulation a method of modeling chance behavior that accurately mimics the situation being considered.

Skewed a unimodal, asymmetric, distribution that tends to slant—most of the data are clustered on one side of the distribution and "tails" off on the other side.

Standard deviation of a binomial random variable X
$$\sigma_x = \sqrt{np(1-p)}.$$

Standard deviation of a discrete random variable X
$$\sigma_x = \sqrt{\sigma_x^2}.$$

Standard deviation used to measure variability of a data set. It is calculated as the square root of the variance of a set of data, $s = \sqrt{\dfrac{\sum(x_i - \overline{x})^2}{n-1}}$.

Standard error an estimate of the standard deviation of the sampling distribution of a statistic.

Standard normal probabilities the probabilities calculated from values of the standard normal distribution.

Standardized score the number of standard deviations an observation lies from the mean, $z = \dfrac{\text{observation} - \text{mean}}{\text{standard deviation}}$.

Statistically significant when a sample statistic is shown to be far from a hypothesized parameter. When the P-value is less than the significance level.

Stemplot also called a stem-and-leaf plot. Data are separated into a stem and a leaf by place value and organized in the form of a histogram.

Strata subgroups of a population that are similar or homogenous.

Stratification part of the sampling process where units of the study are separated into strata.

Stratified random sample a sample in which simple random samples are selected from each of several homogenous subgroups of the population, known as strata.

Subjects individuals in an experiment that are people.

Symmetric the distribution that resembles a mirror image on either side of the center.

Systematic random sample a sample where every kth individual is selected from a list or queue.

Test statistic the number of standard deviations (standard errors) that a sample statistic lies from a hypothesized population parameter.

Third quartile symbolized Q_3, represents the median of the upper 50% of a data set.

Transformation changing the values of a data set using a mathematical operation.

Treatments combinations of different levels of the factors in an experiment.

Two-way table a frequency table that displays two categorical variables.

Type I error rejecting a null hypothesis when it is in fact true.

Type II error failing to reject a null hypothesis when it is in fact false.

Undercoverage when some individuals of a population are not included in the sampling process.

Uniform all data values in the distribution have the similar frequencies.

Unimodal a distribution with a single, clearly defined, peak.

Univariate one-variable data.

Variables characteristics of the individuals under study.

Variability the spread in a data set.

Variance used to measure variability, the average of the squared deviations from the mean,
$$s_x^2 = \frac{\sum (x_i - \bar{x})^2}{n-1}.$$

Variance of a binomial random variable X
$$\sigma_x^2 = np(1-p).$$

Variance of a discrete random variable X
$$\sigma_x^2 = \sum_{i=1}^{n} (x_i - \mu_x)^2 \cdot P(x_i).$$

Venn diagram graphical representation of sets or outcomes and how they intersect.

Voluntary response bias bias due to the manner in which people choose to respond to voluntary surveys.

Voluntary response sample composed of individuals who choose to respond to a survey because of interest in the subject.

z-score see standardized score.

Index

A

addition rule, for probability, 105
alpha level, 199
alternative hypothesis, in
    significance tests, 197–198
anticipating patterns, 103–172

B

back-to-back stemplots, 49–52
bar chart, 26, 72–73
bell-shaped graphs, 26
bias, 90
bimodal graphs, 26
binomial distribution, normal
    approximation of, 144
binomial random variable,
    114–119
bivariate data exploration, 52–70
blinding control, in
    well-designed experiment
    characteristics, 96
blocks, 98
box-and-whisker plot, 26
boxplot construction, 43–46

C

categorical data exploration,
    70–81
categorical variables, 25
categorical/count data inference,
    220–221
census, in data collection,
    83–85
central limit theorem, 149–151
chance device, 124
chi-square distribution,
    168–170
chi-square test, 221
    of goodness of fit, 221–225
    of homogeneity, 226–234
    of independence, 235–239

clusters, 87
completely randomized
    design, 98
conditional frequencies, 77
conditional probability, 106
confidence intervals, 173–174
    confidence intervals duality,
        and hypothesis tests,
        219–220
    and confidence level
        interpretation, 174–175
    for the difference between
        two means, 189–192
    for the difference between
        two proportions, 180–182
    general formula for, 176
    for means, 183–185
    for means with paired data,
        186–188
    for proportions, 176–180
    for regression line slope,
        193–196
confidence intervals inference,
    general procedure,
    176–177
confidence level and confidence
    intervals interpretation,
    174–175
confounding, 96
contingency table, 74–81
continuous discrete random
    variables, 110–114
continuous quantitative
    variables, 25
continuous random variables,
    probability distributions
    of, 132
control principle, in
    well-designed experiment
    characteristics, 96–97
convenience sample, 85
correlation, 57–59
critical value, 175

cumulative frequency graph, 26
cumulative relative frequency
    graph, 36

D

data collection methods
    experiments planning and
        conduction, 94–99
    overview, 83–85
    sampling and experimental
        methods in, 83–99
    survey planning and
        conduction, 85–93
data exploration, 25–81
density curve, 132
dependent events, 107
dependent random
    variables, 130
discrete quantitative
    variables, 25
discrete random variables,
    110–114
dotplot method, 26, 28–29,
    49–52
double-blind, 96

E

estimation, in statistical
    inference, 173–196. *See also
    individual entries*
experiments
    in data collection, 83–85
    design of, 98–99
    planning and conduction,
        in data collection
        methods, 94–99
explanatory variable, in bivariate
    data, 52
exponential model,
    in logarithmic
    transformation, 67

F

factors, 94
frequency table, 70–71

G

gaps in graphs, 28
geometric random variable, 119–124
graphical display, spread interpretation, 27
graphs. *See individual entries*

H

histogram display, 29–30
hypothesis tests
    and confidence intervals duality, 219–220
    decision errors in, 200–202
    for difference between two means, 214–219
    for difference between two proportions, 206–209
    hypothesis tests inference, general procedure, 202–203
    for means, 210–213
    for means with paired data, 213–215
    for proportions, 203–206
    for regression line slope, 240–244

I

independent events, 107
independent random variables, 129–132
independent sample means, 152–156
independent sample proportions, 152–153
influential points, 64–66
interquartile range, 27
inunivariate data, 39–44

J

joint frequencies, 75

L

law of large numbers, 104
least-squares regression line, 59–64
linear models, for transformed data, 67–70
linear regression, 59–63
linear transformations, of random variables, 128–129
logarithmic transformation, 67–70
LSRL. *See* least-squares regression line
lurking variable, 59, 96

M

margin of error, 174
marginal frequencies, 75
matched pairs, 98
matched-pairs *t*-test, 213
mean, in univariate data, 37–39
means, confidence intervals for, 183–185
    confidence intervals for difference between two means, 189–192
    means with paired data, confidence intervals for, 186–188
measures of center, mean and median as, 37–39
measures of position, 46–47
measures of variability, 39–43
median, in univariate data, 37–39
mound-shaped graphs, 26–27
multiplication rule, 108
multistage sample, 87

N

nonresponse bias, 91
normal approximation, of binomial distribution, 144
normal distribution, 133–144
    properties, 133–143
null hypothesis, in significance tests, 197

O

observational study, in data collection, 83–85
one-way table, 71
outliers, 64–66
    in graphs, 28
    in univariate data, 39–43

P

parallel boxplots, 49–52
percentiles, 46–47
placebo, 94, 96–97
point and interval estimates, 173
population parameter, 85
power model, in logarithmic transformation, 67
power of significance tests, 201–202
power transformation, 67–70
probability distributions, 103–109
    of continuous random variables, 132
P-value, 198–200

Q

quantitative data
    constructing displays, 28–33
    distributions description, 26
quantitative variables, 25
quartiles, 46–47

R

random digit table, 124
    sampling of, 92–93
random phenomena, 103, 123–128
random sampling, 86, 91
randomization, in well-designed experiment characteristics, 96–97
randomized block design, 98
range, in univariate data, 39–43
regression line slope, confidence intervals for, 193–196